EPIGENETICS AND RESPONSIBILITY

Ethical Perspectives

Edited by
Emma Moormann, Anna Smajdor
and Daniela Cutas

BRISTOL
UNIVERSITY
PRESS

First published in Great Britain in 2024 by

Bristol University Press
University of Bristol
1–9 Old Park Hill
Bristol
BS2 8BB
UK
t: +44 (0)117 374 6645
e: bup-info@bristol.ac.uk

Details of international sales and distribution partners are available at bristoluniversitypress.co.uk

British Library Cataloguing in Publication Data
A catalogue record for this book is available from the British Library

ISBN 978-1-5292-2542-6 hardcover
ISBN 978-1-5292-2543-3 ePub
ISBN 978-1-5292-2544-0 ePdf

Cover design: Hayes Design and Advertising
Front cover image: 'Passing' by Deborah Forsyth, 2016
(https://debbyforsyth.weebly.com/)
Bristol University Press uses environmentally responsible print partners.
Printed and bound in Great Britain by CPI Group (UK) Ltd, Croydon, CR0 4YY.

FSC
www.fsc.org
MIX
Paper | Supporting
responsible forestry
FSC® C013604

Contents

Notes on Contributors

Eman Ahmed is Lecturer in the Faculty of Medicine at Suez Canal University, Egypt. Her current research interests lie within global health ethics, data sharing and clinical ethics including conceptual meanings of health and disease especially in the context of mental conditions. Ahmed's publications include co-authored articles in *Frontiers of Genetics*, *American Journal of Bioethics* and *AJOB Neuroscience*.

Luca Chiapperino is Swiss National Science Foundation (SNSF) Ambizione Lecturer at the STS Lab, Institute of Social Sciences, University of Lausanne, Switzerland. His research interests lie at the intersection of science and technology studies (STS) and public health ethics, with a specific focus on the socio-political and policy dimensions of biosocial themes in post-genomic biosciences. His work has been published in journals such as *Biosocieties*, *Sociology of Health and Illness*, *Journal of Responsible Innovation*, *Social Studies of Science*, *Clinical Epigenetics* and *Nature Medicine*.

Daniela Cutas is Associate Professor of Medical Ethics at Lund University, Sweden. Her research interests include reproductive and family ethics and the ethics of research. Her work in these areas has been published in journals such as *Bioethics*, *Health Care Analysis*, *Journal of Medical Ethics* and *Journal of Applied Philosophy*. She is the principal investigator of a research project on 'Reproducing the Family: An Ethical Analysis of Intra-familial Access to Reproductive Potential', funded by the Marcus and Amalia Wallenberg Foundation. She is a co-editor of the volumes *Families: Beyond the Nuclear Ideal* (Bloomsbury Academic, 2012) and *Parental Responsibility in the Context of Neuroscience and Genetics* (Springer, 2017).

Maria Hedlund is Associate Professor of Political Science at Lund University, Sweden. She is interested in questions on responsibility and legitimacy in the relationship between expertise and democracy, particularly related to policy making on emerging technologies. Her publications include articles on ethics expertise in *Critical Policy Studies* and *Res Publica*, on distribution of responsibility for AI development in *AI in Society*, *AI and Ethics* and *Frontiers* *in Human Dynamics*, on epigenetic responsibility in *Medicine Studies*, and on genetics and democracy in *Journal of Community Genetics*.

Kristien Hens is Associate Professor of Bioethics at the University of Antwerp, Belgium. She studies the normative implications of concepts of biology and the ethics of developmental diversity, and has a particular interest in the overlap between environmental ethics and health ethics. She is the principal investigator of the NeuroEpigenEthis project, which was funded by a European Research Council (ERC) Starting Grant. She is the author of *Towards an Ethics of Autism* (Open Book Publishers, 2021) and *Chance Encounters* (Open Book Publishers, 2022).

Emma Moormann is Researcher at the University of Antwerp, Belgium, working in the NeuroEpigenEthics project. In the summer of 2023, she defended her dissertation 'Epigenetics and Moral Responsibilities for Health: A Philosophical Exploration'. Her research interests include responsibilities towards future generations and relationships between the biological and the political. In her work, she combines bioethics, feminist philosophy, philosophy of education, experimental philosophy and interdisciplinary research.

Martin Sand is Assistant Professor of Ethics and Philosophy of Technology at TU Delft, The Netherlands. His work focuses on the status and value of digital utopias and the notion of moral progress in engineering ethics education. His work has been published in titles such as *Journal of Medical Ethics*, *Bioethics*, *Journal of Value Inquiry* and *Journal of Applied Philosophy*.

Anna Smajdor is Professor of Practical Philosophy at the University of Oslo, Norway. She is interested in concepts of medical need, disease and harm, and their intersection with social and scientific norms, especially in the context of reproduction. Her publications include a co-authored volume *From IVF to Immortality: Controversy in the Era of Reproductive Technology* (Oxford University Press, 2007) and the *Oxford Handbook of Medical Ethics and Law* (Oxford University Press, 2021) as well as a range of articles in journals such as *The Cambridge Quarterly of Healthcare Ethics*, *Theoretical Medicine and Bioethics* and *Journal of Medical Ethics*.

Introduction: Epigenetics and Responsibility

Emma Moormann and Kristien Hens

Epigenetics: what is it, and why does it matter?

Epigenetics is a fast-growing field in molecular biology. It studies the ways in which modifications to DNA affect gene expression and cell functioning (Carlberg and Molnár, 2019), providing an interface between the genetic and the environmental. The difference between epigenetics and genetics is located in the prefix 'epi', meaning that epigenetic mechanisms are something upon, attached to, or beyond genetics.[1] Epigenetic information may be regarded as another layer beyond genomic information that not only enriches but also challenges insights from more traditional understandings of genetics. The central 'dogma' of genetics is the idea that there is a one-way progression, whereby the genetic code (DNA) is transcribed into RNA, which is translated into proteins. Epigenetics, however, calls into question the unidirectional assumption of this progression, and shows that the interface between genetics and the environment of the genes is much more complex (Hens, 2022).

By regulating gene expression, epigenetics provides a route for environmental influences, including social factors, to affect the development of phenotypes at a molecular level. Epigenome-wide analysis and similar technologies help us to discover the large-scale molecular alterations caused by environmental influences, ranging from food intake during pregnancy to particulate matter related to pollution (Fazzari and Greally, 2010; Rosen et al, 2018; Mancilla et al, 2020). Although the mechanisms described in the central dogma of genetics remain valid, epigenetics paints a far more intricate picture of human development than has often been assumed in science and the popular media alike. This raises important issues for ethicists and legal scholars. For example, it has been suggested that epigenetic changes may be passed on to future generations, extending the scope of responsibility that people may have towards current or future offspring. Moreover, although many ethicists have reflected on the challenges related to the application of CRISPR/Cas9 (a precise gene-editing technology) in human embryos, changing gene expression may be more feasible than changing the genes themselves. As such, a full appreciation of the impact of epigenetics implies viewing it as a molecular basis for a systemic and plastic concept of human nature, situating humans firmly as dynamically altering and being altered by the systems in which they live (Canguilhem, 2008; Thompson, 2010). Which types of responsibility do people have in light of these new findings? How do such findings influence philosophical conceptions of moral responsibility in general? Questions such as these are of central concern to this volume. By looking at these recent developments in biology that reflect a 'dynamic turn' in thinking about human nature, we aim to enrich normative debates on responsibility.

To obtain a somewhat fuller picture of what epigenetics is and what it is not, some short clarifications and demarcations are necessary. Even though contemporary epigenetics as a research field has existed for no more than three decades, various study domains have already been established. The definitions of those domains may vary, and there is often considerable overlap between them. Environmental epigenetics research investigates the ways in which epigenetic alterations may mediate effects caused by environmental exposures or toxins (Jirtle and Skinner, 2007; Bollati and Baccarelli, 2010; Niewöhner, 2011). Neuroepigenetics concerns the regulation of DNA in the nervous system (Sweatt, 2013). Epigenetic epidemiology combines insights from epigenetics with those from epidemiology to improve our understanding of the mechanisms behind observations of interactions between environmental, genetic and stochastic factors and the distribution of diseases (Jablonka, 2004; Heijmans and Mill, 2012). Finally, it is important to mention epigenomics. This is a field of research that focuses on broad or even genome-wide profiles or patterns of epigenetic modifications and their effects (Kato and Natarajan, 2019). Recent research has also been investigating how epigenomics may fruitfully engage with other 'omics' domains such as genomics, which studies the whole of the genetic material in an organism (the genome), and proteomics, a field dedicated to the large-scale study of proteins (Zaghlool et al, 2020; van Mierlo and Vermeulen, 2021). In STS (Science and Technology Studies) and ELSI (Ethical, Legal and Social Implications) literature on epigenetics, the terms 'epigenetics' and 'epigenomics' are sometimes used interchangeably. Although more can be said about the relationship between the two, for our purposes we consider epigenetics to be the more general term, and epigenomics as a field within epigenetics research that focuses especially on the scale of the epigenome but that may nonetheless be regarded as part of the bigger epigenetic project.

The question of whether the epigenetic marks that a person accumulates from their environment may be transmitted to subsequent generations has been widely discussed over the past two decades. Most epigenetic programming is rewritten or reset between generations, but there is increasing evidence that this is not always the case. When considering the transmission of epigenetic marks (such as histone modifications and DNA methylation patterns) between generations, it is important to distinguish between transgenerational and intergenerational effects. Intergenerational epigenetic inheritance refers to epigenetic marks in offspring that are the result of direct exposure of their germline to environmental stressors. This means that intergenerational inheritance is limited to the first generation of male offspring and the first and second generations of female offspring (Cavalli and Heard, 2019). The second generation of female offspring is included because environmental triggers during pregnancy may directly affect the oocytes (egg cells) that are already present in the fetus. A famous example of intergenerational epigenetic inheritance is the famine during the Dutch Hunger Winter of 1944–1945. The children of mothers who experienced this famine during their pregnancy were six decades later found to have less DNA methylation of the imprinted *IFG2* (Insulin Like Growth Factor 2) gene, which is associated with the risk of metabolic diseases (Heijmans et al, 2008). These and other findings contribute to empirical support for the hypothesis that early-life environmental conditions can cause epigenetic changes in humans that persist throughout their lives and on into the next generation(s) (Heijmans et al, 2008; Painter et al, 2008; Lillycrop, 2011). The public discourse and research are often focused on women, perhaps based on 'implicit assumptions about the "causal primacy" of maternal pregnancy effects' (Sharp et al, 2018, p 20). However, epigenetics offers an opportunity to show how not only influences *in utero*, but also paternal factors and postnatal exposures in later life, play a role in the health of offspring. Thus, in epigenetics research, attention is also paid to paternal effects such as the influence of the father's diet on spermatogenesis and offspring health (Rando, 2012; Milliken-Smith and Potter, 2021; Pascoal et al, 2022). Transgenerational epigenetic inheritance is more contested. It denotes the indirect transmission of epigenetic information that is passed on to gametes without alteration of the DNA sequence (Carlberg and Molnár, 2019). This means that we can only speak of transgenerational inheritance if the epigenetic effects of exposures of the current generation are still present in the second generation of male offspring or the third generation of female offspring (Cavalli and Heard, 2019). So far, most transgenerational epigenetic effects have been discovered in plants and non-human animals such as rats and mice. For example, researchers working with mice have found third-generation epigenetic effects of maternal diet (Dunn and Bale, 2011) as well as social stress levels (Matthews and Phillips, 2012), although others argue that multigenerational inheritance of methylation patterns in mice is an exception rather than the rule (Kazachenka et al, 2018). A study of *Caenorhabditis elegans* worms by Klosin and colleagues also had impressive results (Klosin et al, 2017). They genetically modified these worms to glow when exposed to a warm environment. Not only did the worms start to glow more when the temperature was raised, but they also retained their intense glow when researchers lowered the temperature again. Moreover, 'their progeny inherited the glow and even seven generations further down the line, glowing worms were born. If five generations of *C. elegans* worms were kept in a warm environment, this characteristic was passed on to fourteen generations' (Hens, 2022, p 48).

Such findings in animal research sometimes lead to premature conclusions about human health and disease (Juengst et al, 2014). However, it is virtually impossible in research on human inheritance to exclude potential confounding elements such as changes *in utero* and postnatal effects (Cavalli and Heard, 2019). It is hard to distinguish between 'real' epigenetic inheritance and cases where the offspring are simply exposed to the same experiences or health problems as their parents because the context is reconstructed or culturally inherited. However, there are some studies that suggest that transgenerational epigenetic inheritance is possible, albeit limited, in humans. First, when studying historical data of cohorts in Överkalix, Sweden, researchers found associations between grandpaternal food supply and the mortality rate of their children and grandchildren (Kaati et al, 2002). However, because no molecular data were available, no epigenetic links could be proven. Pembrey and colleagues build on those findings to find evidence of sex-specific male transgenerational inheritance in humans (Pembrey et al, 2006). In a longitudinal study in an area around Bristol, UK, they found transgenerational effects of smoking before puberty on the growth of future male offspring of men. Specifically, early paternal smoking (before puberty) was associated with a greater body mass index in their sons. The researchers posit DNA methylation as a potential mechanism behind the links between acquired epigenetic traits of a generation and the epigenetic marks present in the next generations.

Epigenetics also relates to research into the developmental origins of health and disease, or DOHaD, which may be defined as the study of how the early-life environment affects the risk of diseases from childhood to adulthood (Bianco-Miotto et al, 2017). DOHaD also studies the mechanisms involved, which means that there are intricate connections between DOHaD and epigenetics (Vickers, 2014). A core assumption of DOHaD is humorously summarized by Maurizio Meloni as 'We are not so much what we eat, but what our parents ate' (Meloni, 2016, p 209). Thus, both fields overlap, but only partly: epigenetics has a broader focus than just prenatal and perinatal exposures, whereas DOHaD also studies other mechanisms than epigenetic alterations. Many epigeneticists, especially those working in fields such as environmental epigenetics and 'social epigenomics', also see their work as a contribution to the body of knowledge on social determinants of health. These are conditions in people's social and physical environments that influence health outcomes throughout their life course (Mancilla et al, 2020). Such conditions may include the influence of one's family and neighbourhood and one's broader social context, as well as values, attitudes, knowledge and behaviours (Notterman and Mitchell, 2015). Mancilla and colleagues, for example, argue that epigenetics is not the only field that can shed light on social determinants of health, but one that can contribute to explanations of the ways in which socio-environmental factors influence our biology through epigenetic modifications (Mancilla et al, 2020).

The role of the epigeneticist then lies primarily in discovering more about the mechanisms that connect environmental triggers to gene expression (Milliken-Smith and Potter, 2021). A well-known example of such research is the work of McGuinness and colleagues, who investigated the relationship between socio-economic and lifestyle factors and epigenetic profiles in Glasgow, UK, a city that is known for its socio-economic and health disparities. The data were gathered in the context of a broader study on the psychological, social and biological determinants of ill health (pSoBid)[2]. They found lower levels of global DNA methylation in those with a low socio-economic status as well as participants who did manual work. Lower global DNA methylation content was in turn associated with biomarkers of cardiovascular diseases and inflammation (McGuinness et al, 2012). As Milliken-Smith and Potter note, we must be aware that the dynamic between social processes and (epi)genetic information about our health goes two ways. Authors such as McGuiness and colleagues primarily focus on providing 'an explanatory link between the social determinants of health and physiological outcomes'. However, 'a critical appraisal of how this emerging epigenetics knowledge is debated and employed' can highlight how existing biases and disparities may sometimes be reinforced in the social determinants of health framework (Milliken-Smith and Potter, 2021, p 1). We would like to add that researchers, especially those working on the ethical aspects of epigenetics, may benefit from using an intersectional approach that is sensitive to the interplay between various social and environmental conditions (Collins and Bilge, 2016).

Furthermore, epigenetics has shed some new light on our understanding of the development of diseases and disabilities. In the following paragraphs, we give some examples of conditions that are being researched by epigeneticists. It is worth noting that some of the health issues mentioned here, such as stress and obesity, have been posited as both causal contributors to disease development and the outcome of epigenetic processes.

Exposure to stress in the womb or during early childhood has been associated with epigenetically mediated adverse health effects. For example, childhood maltreatment may trigger long-lasting epigenetic marks, contributing to post-traumatic stress disorder in adult life (Mehta et al, 2013). Epigenetic studies have also found that stress in early life can contribute to behaviour that is typical of attention-deficit hyperactivity disorder (Bock et al, 2017). Additionally, Oberlander and colleagues found that the methylation status of the human *NR3C1* (nuclear receptor subfamily 3 group C member 1) gene in newborns is sensitive to maternal depression. They argue that these findings suggest a potential epigenetic process that links the antenatal mood of the mother to the ways that infants respond to new situations, such as an increased stress response to new visual stimuli (Oberlander et al, 2008).

Pollution has numerous harmful effects on health. Emerging data indicate that exposure to air pollution brings about epigenetic changes. These changes may in turn influence inflammation risk and exacerbate the risk of developing lung diseases (Rider and Carlsten 2019). It is well known that lead is a common neurotoxic pollutant that disproportionally affects the health of children. Evidence for the epigenetic basis of the effects of lead is increasing (Senut et al, 2012; Wang et al, 2020).

The epigenetic mechanisms behind the development of metabolic conditions are becoming well-documented. Molecular links between environmental factors and type 2 diabetes have been discovered (Ling and Groop, 2009; Slomko et al, 2012; Rosen et al, 2018), as well as mechanisms that regulate the expression of genes associated with diabetic kidney disease (Kato and Natarajan, 2019). Various studies have also looked into the epigenetics behind obesity, both as a contributory factor and as a health outcome (Lillycrop, 2011; Slomko et al, 2012; Rosen et al, 2018). As type 2 diabetes patients are often more likely to suffer from cardiovascular disease, the influence of environmental factors and the diet of ancestors on the epigenome has also been investigated (Kaati et al, 2002; Lillycrop, 2011). Like stress, obesity has been posited not merely as a health outcome but also as a causal factor that induces other epigenetically mediated conditions. For example, there seems to be an association between overweight in prepubescent boys and diminished lung function and asthma in those boys' adult offspring (Lønnebotn et al, 2022).

Neuroepigeneticists investigate the crucial role that epigenetic regulation plays in the development and functioning of the brain. Conditions for which epigenetic regulatory mechanisms have been suggested include Parkinson's disease, Huntington's disease, schizophrenia, epilepsy, Rett syndrome and depression (Tsankova et al, 2007; Carlberg and Molnár, 2019). Much research is geared towards a better aetiological understanding of neurodevelopmental conditions such as Tourette's syndrome (Müller-Vahl et al, 2017), attention-deficit hyperactivity disorder (Bock et al, 2017; Pineda-Cirera et al, 2019; Wang and Jiang, 2022) and autism (Schanen, 2006; Eshraghi et al, 2018; Waye and Cheng,2018; Gowda and Srinivasan, 2022; Wang and Jiang, 2022). However, there is still much uncertainty about the concrete causal evidence that may be implicated in the development of such conditions (Wang and Jiang, 2022).

In addition to offering new understandings into the ways in which specific diseases arise, epigenetics may also suggest new routes for therapy. Epigenetic changes appear to be more readily reversible than genetic ones (Hens, 2022). This reversibility holds potential for epigenetic therapy, as epigenetic marks such as methylation patterns may be seen as targets for medical interventions and treatments (Heerboth et al, 2014; Carlberg and Molnár, 2019; Nakamura et al, 2021). Many of the clinical research efforts in this domain are directed toward treatments of cancers (Falahi et al, 2014; Lu et al, 2020). Cancer cells are often characterized by epigenetic drifts: the divergence of the epigenome as a function of age due to stochastic changes in methylation (Shah et al, 2014). Many tumours are associated with epigenetic reprogramming (Carlberg and Molnár, 2019). While some studies have investigated the possibility of epigenetic interventions in general, others focused on specific types of cancer such as breast cancer (Falahi et al, 2014) and prostate cancer (Pacheco et al, 2021). Lu and colleagues list so-called 'epidrugs' in clinical trial, with targets also including melanoma, lymphoma, ovarian cancer, bladder cancer and brain tumours (Lu et al, 2020). Research on epidrugs for other conditions is also prolific. Recent projects have aimed at targeting conditions such as COVID-19 (Zannella et al, 2021), hypercholesterolaemia (Paez et al, 2020), neurodegenerative diseases (Janowski et al, 2021), autoimmune diseases such as chronic kidney disease (Tejedor-Santamaria et al, 2022), and depression (Tsankova et al, 2007).

Epigenetics: old wine in new bottles?

Do all these advances in epigenetic knowledge suggest that there is something scientifically or ethically unique about epigenetics to such a degree that we should dedicate an entire volume to it? After all, thousands of books and papers have already been written about genetics and its ethical implications. Is epigenetic exceptionalism – a term coined by Mark Rothstein in line with Thomas Murray's 'genetic exceptionalism' – warranted (Murray, 2019; Rothstein, 2013)? In other words, are new findings in epigenetics so 'extraordinary in kind or degree' that they necessitate new analytical frameworks or novel approaches to deal with their unique character (Rothstein, 2013, p 733)? Before discussing answers to this question, a distinction must be drawn between the potential revolutionary scientific character of findings in epigenetics on the one hand, and the potential unique ethical and social implications of such findings, including those with regard to responsibility, on the other. Rothstein argues that the label of scientific epigenetic exceptionalism is warranted on at least five grounds. First, he contends that epigenetic changes occur much more frequently than mutations in DNA sequences. Moreover, 'an individual's susceptibility to epigenetic change is highly dependent on the dose of the environmental agent and the stage of development at which exposure occurs' (Rothstein, 2013, p 734). Furthermore, he notes that epigenetic changes are intrinsically reversible and tissue- and species-specific. He concludes: 'From a scientific standpoint, epigenetic discoveries are extraordinarily exciting because they represent a new way of understanding the processes by which various harmful exposures cause disease in humans and, in some cases, their offspring. Furthermore, epigenetics could point the way to new methods of preventing and treating numerous disorders' (Rothstein, 2013, p 734).

Laura Benítez-Cojulún discusses the use of various terms used by researchers in describing the significance of epigenetics. Some researchers talk of epigenetics as evoking 'a substantial transformation' (Benítez-Cojulún, 2018, p 135), others use the terms 'epigenetics revolution' (Meloni, 2015, p 141), 'epigenetic turn' (Nicolosi and Ruivenkamp, 2012, p 309) or 'epigenetic shift' (Willer, 2010, p 13). Some use the less favourable term 'epigenetics hype' (Maderspacher, 2010; Deichmann, 2016) to describe 'the far-reaching, revolutionary claims of having discovered entirely new mechanisms of heredity and evolution which are supposed to replace older concepts' (Deichmann, 2016, p 252). Juengst and colleagues appear to consider that exceptionalist language itself is what makes epigenetics exceptional, noting that 'scientific hyperbole rarely generates the level of professional and personal prescriptions for health behaviour that we are now seeing in epigenetics' (Juengst et al, 2014, p 427). Based on a series of in-depth interviews, Kasia Tolwinski has shown that scientists working on epigenetics hold a variety of views with regard to the impact and future of their field. She notes that some epigeneticists are 'champions' of epigenetics as a promising new field. In contrast, others hold a more moderate position, and still others may be considered 'sceptics' regarding the novelty or autonomy of epigenetics as a discipline (Tolwinski, 2013).

The ethical and social implications of epigenetics findings depend partly on their perceived scientific status. However, arguing for some kind of scientific exceptionalism does not necessarily commit one to the view that ethical implications are equally exceptional. Rothstein, for example, does not think that the scientifically distinctive features of epigenetics warrant an ethical exceptionalist approach, stating that 'there is nothing inherently unique about the science of epigenetics that it demands an entirely new ethical paradigm and legal regime' (Rothstein, 2013, p 734). Researchers interviewed by Martyn Pickersgill generally hold similar positions. They 'expressed various kinds of unease about the notion that epigenetic research held straightforward implications for healthcare and society' (Pickersgill, 2021, p 609). Moreover, the respondents 'did not generally conclude that there were immediate ethical ramifications distinct to epigenetics' (p 610).

Jonathan Huang and Nicholas King agree. They do not wish to 'shy away from the potential of epigenetic research' (Meloni and Testa, 2014, p 129). They believe it 'holds promise in identifying and clarifying the different ways in which environments, broadly construed, directly interact with human biology, both within and across generations' (Huang and King, 2018, p 77). However, they have a few concerns. First, they note that 'there is already copious evidence for the impact of social, economic and environmental factors on the health of current and future generations' (p 75). Additionally, they point out that 'epigenetic mechanisms do not in themselves necessarily produce disadvantage; they always work in concert with extant social and economic disadvantages. As such, the injustice of a particular epigenetic variation is always perfectly circumscribed by an existing mechanism of disadvantage, which includes both a prior recognition of a disadvantaged group and an undesirable outcome' (p 74). With regards to responsibility theories in particular, they believe that commentators should refrain from the impossible enterprise of ascribing responsibility and remedy based on epigenetic findings alone, because such findings 'never imply who should be held responsible for any particular causal mechanism' (p 73). They conclude that, in many instances, 'the role of epigenetics is to recapitulate existing claims rather than generate new ones' (p 78). Moreover, they warn against straightforwardly 'using epigenetics to bolster existing ethical claims' (p 73) because of the difficulties involved in characterizing epigenetic changes as harmful and in 'separating unjust epigenetic variations from the social or environmental processes that produced them' (p 73).

Other authors, such as Maria Hedlund, a contributor to this volume, lean more towards the idea that at least there should be a 'change in degree' (Hedlund, 2012) in the ethical response to new findings in epigenetics. She argues that certain ethical concepts or themes, such as collective responsibility, should be used more. Luca Chiapperino, also a contributor to this volume, holds that 'epigenetics poses no new ethical issue over and above those discussed in relation to genetics' (Chiapperino, 2018, p 49). However, he does believe that epigenetics may have important implications for pre-existing ethical issues, arguing that 'epigenetics encourages … "thickening" moral exercises of privacy, responsibility, justice and equity with a complex biosocial description of situations, of persons or actions, in order to afford their significantly balanced evaluation' (p 59). Findings in epigenetics urge us to 'adjust and refine, in a situated manner, the problem frames and categories that inform our ethical and political questions as well as judgements' (p 59).

Similarly, Charles Dupras and Vardit Ravitsky argue that 'the normative accounts of epigenetics do require a heightened degree of bioethical attention, especially considering its potential impact on the political theory of the family and its relation to social as well as intergenerational justice' (Dupras and Ravitsky, 2016, p 2). Rothstein and colleagues argue that most ethical issues related to epigenetics are similar to those already raised by genetics. However, they hold that 'the role of environmental exposures in producing epigenetic effects adds new concerns' such as those about individual and societal responsibilities to prevent hazardous exposures and the multigenerational impact of such exposures (Rothstein et al, 2009, p 2).

Responsibility: a complex relationship

When considering fairness and justice issues in public health, the concept of responsibility has often proven to be an indispensable tool. Epigenetics scholarship is no exception. It seems safe to say that issues related to responsibility are the most-discussed ones in the context of the ethics of epigenetics. A growing body of literature exists on responsibility for actions, such as causing or avoiding epigenetic harm or health and damaging or protecting one's epigenome. The current volume builds on this literature. To do this, let us first investigate what is meant by the concept of 'responsibility'. 'Responsibility' has a wide variety of meanings that are often highly context-dependent. Philosophers of action, ethicists and legal scholars alike have developed competing but often overlapping taxonomies of kinds of responsibility. Here, we introduce the reader to a few general distinctions that will return in other chapters of this volume.

Questions regarding normative responsibility typically involve an analysis of three aspects: who (1) is responsible for what (2) concerning whom (3)? Additionally, we may ask based on which normative standard (4) we wish to hold an agent responsible (Neuhäuser 2014).

Who

The subject of responsibility may be an individual agent, a group of individuals, or a collective agent. The idea that it makes sense to ascribe responsibility to individuals goes relatively unchallenged (a notable exception being Waller, 2011). Although debates about the requirements for and limitations to individual responsibility ascriptions are central to the philosophy of action, the individual agent is often seen as the 'basic bearer of responsibility' (Narveson, 2002). Also not very controversial is the idea of shared responsibility, which is a distributable responsibility that falls on multiple individual agents without them necessarily having any connection or means of communication between them. With collective responsibility, however, matters are more complex. According to proponents of collective responsibility, the collectivity of the subject lies in some qualities of the actions and capacities of the agent that make it appropriate to ascribe responsibility to this collective agent rather than to the individual agents that constitute it. This claim is contested. Methodological individualists do not believe that genuinely collective agents exist. Normative individualists argue that, even if they do, it would be wrong to ascribe responsibilities to them rather than their individual members (Smiley, 2022). Most of the contributions in this volume, however, consider that collective responsibility is a philosophically sound and ethically fruitful concept in the context of epigenetics.

What

Agents may be held responsible for a variety of situations, outcomes, tasks or actions. These are the objects of responsibility. As many authors in this volume point out, responsibility claims are either forward- or backward-looking. Backward-looking approaches, often focusing on whether an agent deserves praise or blame for a specific state of affairs, are the most common in philosophical work on responsibility. Conversely, what is specific about the lesser-discussed forward-looking responsibilities is that they are 'ascribed for the purpose of ensuring the success of a particular moral project rather than for the purpose of gauging the moral agency of a particular group' (Smiley, 2014, p 6). A particularly salient aspect of epigenetics in this regard seems to be the potential reversibility of epigenetic changes (Falahi et al, 2014). Does such reversibility relieve people or collectives of part of their forward-looking responsibility? Do we invest in restorative strategies rather than preventive strategies, or do we invest in both? Most authors in this volume consider both kinds of responsibility relevant to epigenetic responsibility debates. Chapter 2 draws upon work by Linda Radzik (2014) to introduce an additional distinction between the orientation and justification of responsibility ascriptions.

Whom

Generally, when agents have specific responsibilities, these responsibilities are focused on another agent or group of agents. For example, a corporation may be responsible for limiting its environmental impact because the inhabitants of the neighbourhood close to the factory grounds have suffered from its activities. In the context of epigenetics, scholars often urge us all to consider the epigenetic responsibilities we may have towards our offspring and future generations in general (Chiapperino, 2018). Environmental influences on gene expression may affect future children during pregnancy and before people even consider having children. Does knowledge of epigenetic heritability increase individual responsibility, or is there a heightened collective responsibility to ensure a healthy environment for procreation over a lifetime? The potential heritability over generations of epigenetic changes complicates the issue further. Should people change their behaviour if their activities may affect the health of their grandchildren or great-grandchildren?

Should this fact be part of policy decisions? Does it even make sense to say that people who do not yet exist, or who might never come into existence, are the indirect object of a responsibility relationship? Various philosophers have pointed out that, when the people who are impacted by our choices do not yet exist, this may seriously complicate our moral reasoning about those choices. It perhaps comes as no surprise that Derek Parfit's famous 'non-identity problem', which arises from the tension between those complications and our intuitions (Parfit, 1984), has been the focus of various authors working on the ethics of epigenetics (for example Räsänen and Smajdor, 2022; Chapter 4 of this volume).

Basis

Responsibility may have a variety of normative standards, such as moral, causal, legal or political ones. Although it is sometimes very hard to draw the line between those kinds of responsibility in practice, this volume engages primarily with debates about moral responsibility in the context of epigenetics.

Epigenetic responsibility

Whether we conceive of human biology as something static and separate from environmental influence or as dynamic, in constant interaction with, influencing and influenced by the environment, has implications for our understanding of responsibility. For example, concepts of human nature play a role in the debate on what to do about environmental change and who should do it. It has been suggested that humans could be genetically engineered to mitigate or adapt to harmful environmental changes and reduce carbon emissions (Liao et al, 2012). Thus, changing ourselves could be a response to the problems we face concerning the environment. Such suggestions look to genetics to solve global problems in ways that may seem unjustifiably optimistic. At the same time, appeals to human nature are sometimes used as arguments against the acceptability of specific technologies. For example, Fukuyama has argued that human nature, as 'the sum of the behaviour and characteristics that are typical of the human species, arising from genetic rather than environmental factors', is a guiding principle and that any genetic technologies would unacceptably change human nature (Fukuyama, 2003, p 130) – as such, using or subsidizing these technologies is regarded as irresponsible. Interestingly, it appears that both those who argue in favour of modifying humans to adapt to the environment or to increase their health and those who are against modification of human nature take one aspect of human nature for granted: that it is genetically determined.

However, geneticists and biologists have always been aware that the unidirectional central dogma of genetics cannot explain certain phenomena. Philosophers of biology have reflected extensively on how plasticity, the ability of organisms to adapt flexibly to environmental change, affects the nature–nurture distinction (West-Eberhard, 1989; Bateson and Gluckman, 2011; Nicoglou, 2011; Baedke, 2019). Findings in epigenetics, as well as other observations in biology, appear to challenge the idea that human norms can be understood apart from an individual's environmental context (Oyama, 2000; Keller, 2010). For example, Griffiths has suggested that human nature results from the whole organism–environment system that supports human development. As such, he challenges the assumption that human nature is something 'from within' (as in a genetic blueprint) or that human nature is universal (Griffiths, 2011). Moreover, as Hens points out in Chapter 1, the concept of nature as distinct from culture or as static may, in itself, be one that is prevalent only in a specific geographically and temporally defined area.

How does a more dynamic view of human nature influence conceptions of moral responsibility? This is the overarching question that concerns the editors and contributors of this volume. By looking at the recent developments in biology that reflect this 'dynamic turn', namely epigenetics and microbiome research, we aim to enrich normative debates on responsibility for health. There has been a particularly lively debate with regard to which kind of responsibility concepts to use when discussing the ethically salient characteristics of epigenetics. This volume builds on such debates and offers new contributions to them.

Overview of the chapters

In Chapter 1, Kristien Hens reflects on the different meanings of epigenetics. She argues that a developmental view of life, as championed by Waddington and others (Waddington, 2012; Jablonka and Lamb, 2014), can help shed light on the role that bioethicists can play in research projects. She draws on the example of autism research to illustrate how bioethicists can work with scientists to challenge reductionist views of life that consider human beings and their challenges as merely the result of either genetic or environmental factors. In such a context, acknowledging the importance of integrating experiences of stakeholders in the research is extremely important.

In Chapter 2, Emma Moormann discusses the concept of 'forward-looking collective responsibility' in ethical debates involving epigenetics. After reviewing previous uses of the concept in an epigenetics context, she goes on to formulate suggestions for the integration of forward-looking collective responsibility in a framework of responsibility for epigenetic justice. Starting from an intersectional feminist, egalitarian perspective, she uses the case of a Mexico City neighbourhood to show how those concerned about epigenetic responsibility can resist calls for 'epigenetic eliminativism', the idea that we should not and perhaps cannot make responsibility claims in light of epigenetic findings.

Luca Chiapperino and Martin Sand also delve into the issue of collective epigenetic responsibilities in Chapter 3. They build on previous work on (moral) luck that questions the causality condition of epigenetic responsibility claims for both individuals and collective agents. They argue that collective agents are subject to the complexities and uncertainties of epigenetic mechanisms that limit their epigenetic knowledge as well as their capacity to act on it. However, they consider it important to identify normative reasons to let collective agents play a role in an effective societal scenario of epigenetic knowledge. Thus, they argue that residual epigenetic responsibilities may be ascribed to collective agents on alternative grounds. Drawing on notions of 'aretaic blame', the authors propose a model for collective commitments to the protection of our epigenomes that is based on evaluation of the worth of these collective agents.

In Chapter 4, Anna Smajdor explores the question of whether epigenetic alterations to sperm, eggs or embryos may be viewed as harmful to resulting offspring. In particular, she addresses the 'non–identity problem', which has been instrumental in shaping the debate in reproductive ethics. She notes that the concept of genetic identity is deeply problematic. Focusing on epigenetics may resolve some of these problems, but in turn raises others.

In Chapter 5, Daniela Cutas analyses the implications of findings in epigenetics for determination of responsibility for children, particularly for parental responsibility. She reviews various accounts of responsibility for children, and shows how these have been based on widely shared assumptions about children being, ultimately, 'made' by their biological (genetic) parents. By blurring the boundary between social and biological contributions to children's lives, epigenetics extends the reach of responsibility for children, and thereby calls into question the proportion of responsibility that should fall on the shoulders of the 'biological' parents. As many of the forces that shape children's lives are systemic rather than individual, remedial action must also be systemic.

In Chapter 6, Maria Hedlund broadens the discussion about epigenetic responsibility to investigate the ways in which developments in artificial intelligence (AI) further complicate questions of epigenetic responsibility. She elucidates some of the complexities in the responsibility equation that arise when AI technology in general, and machine learning in particular, are employed to analyse epigenetic data. She concludes with a call for interdisciplinary collaboration and the need to focus attention on the ethical dimensions of precision medicine.

Chapter 7, by Kristien Hens and Eman Ahmed, goes beyond epigenetics to discuss the microbiome. As with epigenetics, recent findings regarding the microbiome–gut–brain axis challenge atomistic and static conceptions of organisms. The authors investigate how the questions raised by epigenetics are also relevant for ethical questions surrounding the microbiome. They describe the idea of the 'holobiont', and how it matters for responsibility. This raises questions about privacy: what kind of private information can we get from stool samples? Is this different from genetic information? How does the link between the microbiome and mental health affect our self-understanding? They end by suggesting that, even more than epigenetics, microbiome research posits human beings and other organisms as firmly entangled with, and partially defined by, the environment.

Notes

¹ Although we follow many authors who explain the term in this way, we acknowledge that it cannot serve as a proper aetiology of the term. Stotz and Griffiths note that Waddington introduced the term as a fusion of 'epigenesis' and 'genetics', rather than as 'genetics' with the prefix 'epi' (Stotz and Griffiths, 2016).

² See: https://www.gcph.co.uk/publications/421_psychological_social_and_biological_determinants_of_ill_health_psobid

Acknowledgements

This volume is a deliverable of the European Research Council Starting Grant NeuroEpigenEthics (grant number 804881), which investigates concepts of responsibility in the context of epigenetics and neurodevelopmental disorders. The editors are grateful for insightful feedback on a volume draft from colleagues at the Unit of Medical Ethics at Lund University, Sweden: Linus Broström, Göran Hermerén, Kristina Hug, Mats Johansson and Jenny Lindberg. Additionally, they would like to thank NeuroEpigenEthics team members François-Lucien Vulliermet and Lies Van Den Plas for proofreading some of the chapters.

Contributor statement

E.M. and K.H. both contributed to the structure and content of this chapter. E.M. used parts of the first chapter of her doctoral thesis for this chapter. K.H. and E.M. both gave feedback on and edited the whole manuscript and agree with this final version.

Statement on editorship of the volume

E.M., A.S. and D.C. have reviewed all chapters and provided feedback to all contributors on several versions of all chapters. All decisions regarding the organization of the volume have been taken in agreement by the three editors.

References

Baedke, J. (2019) 'O organism, where art thou? Old and new challenges for organism-centered biology', *Journal of the History of Biology*, 52(2): 293–324.

Bateson, P.P.G. and Gluckman, P.D. (2011) *Plasticity, Robustness, Development and Evolution*, Cambridge: Cambridge University Press.

Benítez-Cojulún, L. (2018) 'The history of epigenetics from a sociological perspective', *InterDisciplines*, 9(2): 135–61.

Bianco-Miotto, T., Craig, J.M., Gasser, Y.P., van Dijk, S.J. and Ozanne, S.E. (2017) 'Epigenetics and DOHaD: from basics to birth and beyond', *Journal of Developmental Origins of Health and Disease*, 8(5): 513–19.

Bock, J., Breuer, S., Poeggel, G. and Braun, K. (2017) 'Early life stress induces attention-deficit hyperactivity disorder (ADHD)-like behavioral and brain metabolic dysfunctions: functional imaging of methylphenidate treatment in a novel rodent model', *Brain Structure and Function*, 222(2): 765–80.

Bollati, V. and Baccarelli, A. (2010) 'Environmental epigenetics', *Heredity*, 105(1): 105–12.

Canguilhem, G. (2008) *Knowledge of Life*, New York: Fordham University Press.

Carlberg, C. and Molnár, F. (2019) *Human Epigenetics: How Science Works*, Berlin: Springer.

Cavalli, G. and Heard, E. (2019) 'Advances in epigenetics link genetics to the environment and disease', *Nature*, 571(7766): 489–99. https://doi.org/10.1038/s41586-019-1411-0.

Chiapperino, L. (2018) 'Epigenetics: ethics, politics, biosociality', *British Medical Bulletin*, 128(1): 49–60.

Collins, P.H. and Bilge, S. (2016) *Intersectionality. Key Concepts*, Cambridge: Polity Press.

Deichmann, U. (2016) 'Epigenetics: the origins and evolution of a fashionable topic', *Developmental Biology*, 416(1): 249–54.

Dunn, G.A. and Bale, T.L. (2011) 'Maternal high-fat diet effects on third-generation female body size via the paternal lineage', *Endocrinology*, 152(6): 2228–36.

Dupras, C. and Ravitsky, V. (2016) 'the ambiguous nature of epigenetic responsibility', *Journal of Medical Ethics*, 42(8): 534–41.

Eshraghi, A.A., Liu, G., Kay, S.-I.S., Eshraghi, R.S., Mittal, J., Moshiree, B. and Mittal, R. (2018) 'Epigenetics and autism spectrum disorder: is there a correlation?', *Frontiers in Cellular Neuroscience*, 12: 78.

Falahi, F., van Kruchten, M., Martinet, N., Hospers, G.A.P. and Rots, M.G. (2014) 'Current and upcoming approaches to exploit the reversibility of epigenetic mutations in breast cancer', *Breast Cancer Research*, 16(4): 412. https://doi.org/10.1186/s13058-014-0412-z.

Fazzari, M.J. and Greally, J.M. (2010) 'Introduction to epigenomics and epigenome-wide analysis', *Statistical Methods in Molecular Biology*, 620: 243–65.

Fukuyama, F. (2002) *Our Posthuman Future: Consequences of the Biotechnology Revolution*, New York: Farrar, Straus and Giroux.

Gowda, V.K. and Srinivasan, V.M. (2022) 'A treatable cause of global developmental delay with autism spectrum disorder due to cobalamin related remethylation disorder', *Indian Journal of Pediatrics*, 89(8): 832.

Griffiths, P.E. (2011) 'Our plastic nature', in S. Gissis and E. Jablonka (eds), *Transformations of Lamarckism: From Subtle Fluids to Molecular Biology*, Cambridge, MA: MIT Press, pp 319–30.

Hedlund, M. (2012) 'Epigenetic responsibility', *Medicine Studies*, 3(3): 171–83. https://doi.org/10.1007/s12376-011-0072-6.

Heerboth, S., Lapinska, K., Snyder, N., Leary, M., Rollinson, S. and Sarkar, S. (2014) 'Use of epigenetic drugs in disease: an overview', *Genetics & Epigenetics* 6: 9–19.

Heijmans, B.T. and Mill, J. (2012) 'Commentary: The seven plagues of epigenetic epidemiology', *International Journal of Epidemiology*, 41(1): 74–8.

Heijmans, B.T., Tobi, E.W., Stein, A.D., Putter, H., Blauw, G.J., Susser, E.S., et al (2008) 'Persistent epigenetic differences associated with prenatal exposure to famine in humans', *Proceedings of the National Academy of Sciences USA*, 105(44): 17046–9.

Hens, K. (2022) *Chance Encounters: A Bioethics for a Damaged Planet* (1st edn), Cambridge: Open Book Publishers. https://doi.org/10.11647/OBP.0320.

Huang, J.Y. and King, N.B. (2018) 'Epigenetics changes nothing: what a new scientific field does and does not mean for ethics and social justice', *Public Health Ethics*, 11(1): 69–81.

Jablonka, E. (2004) 'Epigenetic epidemiology', *International Journal of Epidemiology*, 33(5): 929–35.

Jablonka, E. and Lamb, M.J. (2014) *Evolution in Four Dimensions, Revised Edition: Genetic, Epigenetic, Behavioral, and Symbolic Variation in the History of Life*, Cambridge, MA: MIT Press.

Janowski, M., Milewska, M., Zare, P. and Pękowska, A. (2021) 'Chromatin alterations in neurological disorders and strategies of (epi)genome rescue', *Pharmaceuticals*, 14(8): 765. https://doi.org/10.3390/ph14080765.

Jirtle, R.L. and Skinner, M.K. (2007) 'Environmental epigenomics and disease susceptibility', *Nature Reviews Genetics*, 8(4): 253–62. https://doi.org/10.1038/nrg2045.

Juengst, E.T., Fishman, J.R., McGowan, M.L. and Settersten, Jr, R.A. (2014) 'Serving epigenetics before its time', *Trends in Genetics*, 30(10): 427–9.

Kaati, G., Bygren, L. and Edvinsson, S. (2002) 'Cardiovascular and diabetes mortality determined by nutrition during parents' and grandparents' slow growth period', *European Journal of Human Genetics*, 10(11): 682–8. https://doi.org/10.1038/sj.ejhg.5200859.

Kato, M. and Natarajan, R. (2019) 'Epigenetics and epigenomics in diabetic kidney disease and metabolic memory', *Nature Reviews Nephrology* 15(6): 327–45. https://doi.org/10.1038/s41581-019-0135-6.

Kazachenka, A., Bertozzi, T.M., Sjoberg-Herrera, M.K., Walker, N., Gardner, J., Gunning, R., et al (2018) 'Identification, characterization, and heritability of murine metastable epialleles: implications for non-genetic inheritance', *Cell*, 175(5): 1259–71.

Keller, E.F. (2010) *The Mirage of a Space between Nature and Nurture*, Durham, NC: Duke University Press.

Klosin, A., Casas, E., Hidalgo-Carcedo, C., Vavouri, T., and Lehner, B. (2017) 'Transgenerational transmission of environmental information in *C. elegans*', *Science*, 356(6335): 320–3.

Liao, S.M., Sandberg, A. and Roache, R. (2012) 'Human engineering and climate change', *Ethics, Policy & Environment*, 15(2): 206–21. https://doi.org/10.1080/21550085.2012.685574.

Lillycrop, K.A. (2011) 'Effect of maternal diet on the epigenome: implications for human metabolic disease', *Proceedings of the Nutrition Society* 70(1): 64–72. https://doi.org/10.1017/S0029665110004027.

Ling, C. and Groop, L. (2009) 'Epigenetics: a molecular link between environmental factors and type 2 diabetes', *Diabetes*, 58(12): 2718–25. https://doi.org/10.2337/db09-1003.

Lønnebotn, M., Calciano, L., Johannessen, A., Jarvis, D.L., Abramson, M.J., Benediktsdóttir, B., et al (2022) 'Parental prepuberty overweight and offspring lung function', *Nutrients*, 14(7): 1506. https://doi.org/10.3390/nu14071506.

Lu, Y., Chan, Y.-T., Tan, H.-Y., Li, S., Wang, N. and Feng, Y. (2020) 'Epigenetic regulation in human cancer: the potential role of epi-drug in cancer therapy', *Molecular Cancer*, 19(1): 79. https://doi.org/10.1186/s12943-020-01197-3.

Maderspacher, F. (2010) 'Lysenko rising', *Current Biology*, 20(19): R835–7.

Mancilla, V.J., Peeri, N.C., Silzer, T., Basha, R., Felini, M., Jones, H.P., et al (2020) 'Understanding the interplay between health disparities and epigenomics', *Frontiers in Genetics*, 11: 903. https://doi.org/10.3389/fgene.2020.00903.

Matthews, S.G. and Phillips, D.I. (2012) 'Transgenerational inheritance of stress pathology', *Experimental Neurology*, 233(1): 95–101.

McGuinness, D., McGlynn, L.M., Johnson, P.C.D., MacIntyre, A., Batty, G.D., Burns, H., Cavanagh, J., Deans, K.A., Ford, I. and McConnachie, A. (2012) 'Socio-economic status is associated with epigenetic differences in the PSoBid cohort', *International Journal of Epidemiology* 41(1): 151–60.

Mehta, D., Klengel, T., Conneely, K.N., Smith, A.K., Altmann, A., Pace, T.W., et al (2013) 'Childhood maltreatment is associated with distinct genomic and epigenetic profiles in posttraumatic stress disorder', *Proceedings of the National Academy of Sciences USA*, 110(20): 8302–7.

Meloni, M. (2015) 'Epigenetics for the social sciences: justice, embodiment, and inheritance in the postgenomic age', *New Genetics and Society*, 34(2): 125–51.

Meloni, M. (2016) *Political Biology: Science and Social Values in Human Heredity from Eugenics to Epigenetics*, Basingstoke: Palgrave Macmillan.

Meloni, M. and Testa, G. (2014) 'Scrutinizing the epigenetics revolution', *BioSocieties*, 9(4): 431–56.

Milliken-Smith, S. and Potter, C.M. (2021) 'Paternal origins of obesity: emerging evidence for incorporating epigenetic pathways into the social determinants of health framework', *Social Science & Medicine*, 271: 112066. https://doi.org/10.1016/j.socscimed.2018.12.007.

Müller-Vahl, K.R., Loeber, G., Kotsiari, A., Müller-Engling, L. and Frieling, H. (2017) 'Gilles de la Tourette syndrome is associated with hypermethylation of the dopamine D2 receptor gene', *Journal of Psychiatric Research*, 86: 1–8.

Murray, T.H. (2019) 'Is genetic exceptionalism past its sell-by date? On genomic diaries, context, and content', *The American Journal of Bioethics*, 19(1): 13–15. DOI: 10.1080/15265161.2018.1552038.

Nakamura, M., Gao, Y., Dominguez, A.A. and Qi, L.S. (2021) 'CRISPR technologies for precise epigenome editing', *Nature Cell Biology*, 23(1): 11–22. https://doi.org/10.1038/s41556-020-00620-7.

Narveson, J. (2002) 'Collective responsibility', *The Journal of Ethics*, 6(2): 179–98.

Neuhäuser, C. (2014) 'Structural injustice and the distribution of forward-looking responsibility', *Midwest Studies in Philosophy*, 38: 232–51.

Nicoglou, A. (2011) 'Defining the boundaries of development with plasticity', *Biological Theory*, 6(1): 36–47. https://doi.org/10.1007/s13752-011-0003-5.

Nicolosi, G. and Ruivenkamp, G. (2012) 'The epigenetic turn', *Medicine, Health Care and Philosophy*, 15(3): 309–19.

Niewöhner, J. (2011) 'Epigenetics: embedded bodies and the molecularisation of biography and milieu', *BioSocieties*, 6(3): 279–98. https://doi.org/10.1057/biosoc.2011.4.

Notterman, D.A. and Mitchell, C. (2015) 'Epigenetics and understanding the impact of social determinants of health', *Pediatric Clinics of North America*, 62(5): 1227–40. https://doi.org/10.1016/j.pcl.2015.05.012.

Oberlander, T.F., Weinberg, J., Papsdorf, M., Grunau, R., Misri, S. and Devlin, A.M. (2008) 'Prenatal exposure to maternal depression, neonatal methylation of human glucocorticoid receptor gene (NR3C1) and infant cortisol stress responses', *Epigenetics*, 3(2): 97–106. https://doi.org/10.4161/epi.3.2.6034.

Oyama, S. (2000) *The Ontogeny of Information: Developmental Systems and Evolution*, Durham, NC: Duke University Press.

Pacheco, M.B., Camilo, V., Lopes, N., Moreira-Silva, F., Correia, M.P., Henrique, R. and Jerónimo, C. (2021) 'Hydralazine and panobinostat attenuate malignant properties of prostate cancer cell lines', *Pharmaceuticals*, 14(7): 670. https://doi.org/10.3390/ph14070670.

Paez, I., Prado, Y., Ubilla, C.G., Zambrano, T. and Salazar, L.A. (2020) 'Atorvastatin increases the expression of long non-coding RNAs ARSR and CHROME in hypercholesterolemic patients: a pilot study', *Pharmaceuticals*, 13(11): 382. https://doi.org/10.3390/ph13110382.

Painter, R.C., Osmond, C., Gluckman, P., Hanson, M., Phillips, D.I.W. and Roseboom, T.J. (2008) 'Transgenerational effects of prenatal exposure to the Dutch famine on neonatal adiposity and health in later life', *BJOG*, 115(10): 1243–9.

Parfit, D. (1984) *Reasons and Persons*, Oxford: Clarendon Press.

Pascoal, G.F.L., Geraldi, M.V., Maróstica, Jr, M.R. and Ong, T.P. (2022) 'Effect of paternal diet on spermatogenesis and offspring health: focus on epigenetics and interventions with food bioactive compounds', *Nutrients*, 14(10): 2150. https://doi.org/10.3390/nu14102150.

Pembrey, M.E., Bygren, L.O., Kaati, G., Edvinsson, S., Northstone, K., Sjöström M. and Golding, J. (2006) 'Sex-specific, male-line transgenerational responses in humans', *European Journal of Human Genetics*, 14(2): 159–66.

Pickersgill, M. (2021) 'Negotiating novelty: constructing the novel within scientific accounts of epigenetics', *Sociology*, 55(3): 600–18.

Pineda-Cirera, L., Shivalikanjli, A., Cabana-Domínguez, J., Demontis, D., Rajagopal, V.M., Børglum, A.D., et al (2019) 'Exploring genetic variation that influences brain methylation in attention-deficit/hyperactivity disorder', *Translational Psychiatry*, 9(1): 242.

Radzik, L. (2014) 'Historical memory as forward-and backward-looking collective responsibility', *Midwest Studies in Philosophy*, 38: 26–39.

Rando, O.J. (2012) 'Daddy issues: paternal effects on phenotype', *Cell*, 151(4): 702–8. https://doi.org/10.1016/j.cell.2012.10.020.

Räsänen, J. and Smajdor, A. (2022) 'Epigenetics, harm, and identity', *The American Journal of Bioethics*, 22(9): 40–2. https://doi.org/10.1080/15265161.2022.2105424.

Rider, C.F. and Carlsten, C. (2019) 'Air pollution and DNA methylation: effects of exposure in humans', *Clinical Epigenetics*, 11(1): 131.

Rosen, E.D., Kaestner, K.H., Natarajan, R., Patti, M.-E., Sallari, R., Sander M. and Susztak, K. (2018) 'Epigenetics and epigenomics: implications for diabetes and obesity', *Diabetes*, 67(10): 1923–31.

Rothstein, M.A. (2013) 'Legal and ethical implications of epigenetics', in R.L. Jirtle and F.L. Tyson (eds), *Environmental Epigenomics in Health and Disease*, Berlin: Springer, pp 297–308.

Rothstein, M.A., Cai Y. and Marchant, G.E. (2009) 'Ethical implications of epigenetics research', *Nature Reviews Genetics*, 10(4): 224.

Schanen, N.C. (2006) 'Epigenetics of autism spectrum disorders', *Human Molecular Genetics*, 15(Suppl 2): R138–50. https://doi.org/10.1093/hmg/ddl213.

Senut, M.-C., Cingolani, P., Sen, A., Kruger, A., Shaik, A., Hirsch, H., et al (2012) 'Epigenetics of early-life lead exposure and effects on brain development', *Epigenomics*, 4(6): 665–74.

Shah, S., McRae, A.F., Marioni, R.E., Harris, S.E., Gibson, J., Henders, A.K., et al (2014) 'Genetic and environmental exposures constrain epigenetic drift over the human life course', *Genome Research*, 24(11): 1725–33. https://doi.org/10.1101/gr.176933.114.

Sharp, G.C., Lawlor, D.A. and Richardson, S.S. (2018) 'It's the mother!: how assumptions about the causal primacy of maternal effects influence research on the developmental origins of health and disease', *Social Science & Medicine*, 213: 20–7. https://doi.org/10.1016/j.socscimed.2018.07.035.

Slomko, H., Heo, H.J. and Einstein, F.H. (2012) 'Minireview: Epigenetics of obesity and diabetes in humans', *Endocrinology*, 153(3): 1025–30.

Smiley, M. (2014) 'Future-looking collective responsibility', *Midwest Studies In Philosophy*, 38: 1–11. https://doi.org/10.1111/misp.12012.

Smiley, M. (2022) 'Collective responsibility', in E.N. Zalta and U. Nodelman (eds), *The Stanford Encyclopedia of Philosophy* [online], 19 December. Available from: https://plato.stanford.edu/entries/collective-responsibility/ [Accessed 12 April 2023].

Stotz, K. and Griffiths, P. (2016) 'Epigenetics: ambiguities and implications', *History and Philosophy of the Life Sciences*, 38(4): 22. https://doi.org/10.1007/s40656-016-0121-2.

Sweatt, J.D. (2013) 'The emerging field of neuroepigenetics', *Neuron*, 80(3): 624–32.

Tejedor-Santamaria, L., Morgado-Pascual, J.L., Marquez-Exposito, L., Suarez-Alvarez, B., Rodrigues-Diez, R.R., Tejera-Muñoz, A., et al (2022) 'Epigenetic modulation of Gremlin-1/NOTCH pathway in experimental crescentic immune-mediated glomerulonephritis', *Pharmaceuticals*, 15(2): 121. https://doi.org/10.3390/ph15020121.

Thompson, E. (2010) *Mind in Life: Biology, Phenomenology, and the Sciences of Mind*, Cambridge, MA: Harvard University Press.

Tolwinski, K. (2013) 'A new genetics or an epiphenomenon? Variations in the discourse of epigenetics researchers', *New Genetics and Society*, 32(4): 366–84.

Tsankova, N., Renthal, W., Kumar, A. and Nestler, E.J. (2007) 'Epigenetic regulation in psychiatric disorders', *Nature Reviews Neuroscience*, 8(5): 355–67.

van Mierlo, G. and Vermeulen, M. (2021) 'Chromatin proteomics to study epigenetics – challenges and opportunities', *Molecular & Cellular Proteomics*, 20: 100056. https://doi.org/10.1074/mcp.R120.002208.

Vickers, M. (2014) 'Early life nutrition, epigenetics and programming of later life disease', *Nutrients*, 6(6): 2165–78. https://doi.org/10.3390/nu6062165.

Waddington, C.H. (2012) 'The epigenotype: 1942', *International Journal of Epidemiology*, 41(1): 10–13.

Waller, B.N. (2011) *Against Moral Responsibility*, Cambridge, MA: MIT Press.

Wang, S.E. and Jiang, Y.-H. (2022) 'Epigenetic epidemiology of autism and other neurodevelopmental disorders', in K.B. Michels (ed), *Epigenetic Epidemiology* (2nd edn), Berlin: Springer, pp 405–26.

Wang, T., Zhang, J. and Xu, Y. (2020) 'Epigenetic basis of lead-induced neurological disorders', *International Journal of Environmental Research and Public Health*, 17(13): 4878. https://doi.org/10.3390/ijerph17134878.

Waye, M.M.Y. and Cheng, H.Y. (2018) 'Genetics and epigenetics of autism: a review', *Psychiatry and Clinical Neurosciences*, 72(4): 228–44.

West-Eberhard, M.J. (1989) 'Phenotypic plasticity and the origins of diversity', *Annual Review of Ecology and Systematics*, 20(1): 249–78.

Willer, S. (2010) '"Epigenesis" in epigenetics: scientific knowledge, concepts, and words', in A. Barahona, E. Suarez-Díaz and H-J. Rheinberger (eds), *Preprint 392 – The Hereditary Hourglass: Genetics and Epigenetics, 1868–2000*, [online], pp 13–21. Available from: https://www.mpiwg-berlin.mpg.de/sites/default/files/Preprints/P392.pdf [Accessed 13 April 2023].

Zaghlool, S.B., Kühnel, B. , Elhadad, M.A., Kader, S., Halama, A., Thareja, G., Engelke, R., et al (2020) 'Epigenetics meets proteomics in an epigenome-wide association study with circulating blood plasma protein traits', *Nature Communications*, 11(1): 15. https://doi.org/10.1038/s41467-019-13831-w.

Zannella, C., Rinaldi, L., Boccia, G., Chianese, A., Sasso, F.C. , De Caro, F., et al (2021) 'Regulation of m6A methylation as a new therapeutic option against COVID-19', *Pharmaceuticals*, 14(11): 1135. https://doi.org/10.3390/ph14111135.

1

Epigenetics, Bioethics and a Developmental Outlook on Life

Kristien Hens

Introduction

Epigenetics refers to the molecular mechanisms that control gene expression. These mechanisms are closely tied to environmental influences within the cell, the body and the environment. Epigenetics ensues naturally from genetics, the life science that dominated Western science in the 20th century. After all, it has always been known, even by the staunchest genetic determinist, that there must be mechanisms for transcribing and translating specific genes in specific circumstances. In this respect, the claim that an epigenetic approach to life and ethics offers something unheard of exaggerates the truth. Indeed, for a long time, scientists and ethicists have investigated and discussed environmental influences. However, in this chapter, I argue that epigenetics in a Waddingtonian sense urges us to rethink the object and scope of bioethics and the relationship between ethics and science in general. I hope to demonstrate that thinking about the historical meaning of epigenetics and its relationship to the concept of development can teach us something about the role that bioethics could play in biomedical research projects. In order to do so, I draw on my work on the ethics of research in developmental diversity such as autism. First, I look at the meaning of epigenetics and the closely related term epigenesis. I demonstrate that, in addition to providing insights into molecular mechanisms regulating gene expression, a focus on epigenetics also supports a developmental view of life. I then give an overview of existing bioethical reflections on epigenetics. I suggest that a developmental approach may require some fundamental changes in ethical considerations. Rather than focusing on epigenetics as an alternative to genetics as a cause of phenomena and disorders, a developmental approach may imply emphasis on dynamics, context and experiences in normative reflection. To illustrate this, I explore what epigenetics means for research into developmental diversity in general and autism specifically. Finally, I return to the role that bioethicists could play in thinking with and about science, and make some suggestions as to what the role of bioethicists could be in relation to the aims and subjects of responsible science.

What epigenesis and epigenetics have in common is that they presuppose a developmental perspective on life. Epigenesis is a much older word than epigenetics. The term itself was coined in the 19th century by C.V.F. Wolff, although developmental perspectives on organisms have a much longer history, as the term peri-genesis, coined by Aristotle, suggests (Wessel, 2009). Epigenesis is a view of the development of organisms, and is often contrasted with preformation. If you hold a preformationist view on life, you consider that an organism's form is already there at the very start, from conception onwards. The preformationist view is closely linked to the idea of the homunculus, i.e. a tiny but fully formed human that lives inside the sperm and is merely enlarged during the organism's development within the uterus. Many 20th-century geneticists, in emphasizing the importance of genes rather than the environment, may be viewed as preformationists. After all, they seem to share the view that everything a person or an organism can become, minus some minor adaptations related to lifestyle and environment, is already there from the beginning. Today, few would question the environment's role in the development of organisms. However, some common assumptions in bioethics retain a dash of preformationism, such as the discussion around the non-identity problem (see Chapter 4). Several examples that Parfit uses in his famous book *Reasons and Persons* (Parfit, 1984) imply that what matters for identity is numerical identity: what happens at conception. At this point, we become who we are.

Nowadays, preformationist thinking seems to be out of fashion to the point that calling someone a preformationist may even be an insult. At the same time, the idea that there is some point in our developmental history when our identity becomes fixed, is something many can relate to, and that is assumed in many discussions about responsibility to (future) offspring.

The epigenetic landscape is a way to describe two mechanisms: canalization and plasticity. Canalization is the adjustment of developmental pathways to bring about a uniform developmental result despite genetic and environmental variations. Imagine the cell as a little marble rolling through the canals or valleys of the epigenetic landscape. Depending on the landscape (or the environment), it is sent through specific channels, ending up as a specific cell type or with a specific function. But there is also plasticity. Suppose that the landscape shuffles a bit: a minor rearrangement may have little effect on the cell's trajectories because they are canalized. However, if the landscape is completely rearranged, this will significantly affect the phenotype. Hence, plasticity and canalization are not opposites, but work together. Canalized development enables the organism to adapt to different circumstances. Furthermore, adapting to different circumstances implies that the organism is stable enough to withstand complete disruption. Thus, dynamics and stability are two sides of the same coin (Jablonka and Lamb, 2014; Jablonka, 2016).

In describing the epigenetic landscape, Waddington discussed the development of cells in different cell types within the body during phenotype development. He was not suggesting that this represents a general account of the influences of lifestyle and environment on organisms. However, it is easy to see how the ideas of dynamics and stability, plasticity and canalization, can apply to an organism's interaction with its environment throughout its lifetime. For example, Jablonka and colleagues have used the idea of Waddington's epigenetic landscape to provide a way to think about culture. They argue that we can think of the social landscape as a Waddingtonian landscape: a dynamic pattern of life in a particular community where customs and practices become canalized (Tavory et al, 2014; Jablonka, 2016). Moreover, Waddington himself was very aware of the implications of complex systems thinking for science and society at large, as is apparent in his books *The Scientific Attitude* (Waddington, 1948) and *Tools for Thought* (Waddington, 1977). In what follows, I investigate the bioethical implications of a Waddingtonian approach, focusing on autism.

Epigenetics and bioethics: a marriage made in hell?

Bioethicists have discussed the ethical aspects of epigenetics at length. They have asked themselves how, if at all, epigenetics is different from genetics. However, I argue that the relevance of epigenetics is not so much its novelty. Instead, a Waddingtonian approach suggests a different view of organisms than the one that bioethicists may too often have taken for granted. The various meanings of epigenetics described previously have some things in common. Epigenetics, as a molecular mechanism regulating gene expression and as the more general idea of plasticity and canalization in development, gives biology a history. What happens in an organism's life, what it encounters and experiences, resonates in its biology. All interpretations of 'epigenetics' stress development and how organisms come into being, gain form and interact throughout their lives. From an epigenetic point of view, organisms 'lay down a path in walking', to use the words of the enactivist thinker Francisco Varela (Varela et al, 1992). The concept of epigenetics, in all its interpretations, emphasize the entanglement of organisms with their environment. Epigenetic changes occur in response to environmental influences. However, it would be wrong to see the environment and the genome as two different spheres with equal explanatory power. 'The environment' may mean many things: the location of a cell in the body, lifestyle choices such as those related to food and exercise, physical influence such as environmental pollution, and psychological influences such as stress and nurture. All these spheres influence gene expression in distinct ways. Moreover, as is apparent from Waddington's idea of a network of genes, the question is not one of either/or. Genes and various environmental factors, epigenetics and microbiomes all play a role in the complicated workings of the cell. A developmental view of life is hence not so much a challenge to the central dogma of genetics per se as a challenge to views that consider genes to be the primary or exclusive causes for behaviour, traits and diseases. Moreover, as I argue below, it would be a mistake to look at environmental factors as exact causes, in the way that some consider genes to be. If we conceive of organisms as developing through life in response to input from the environment, this introduces an element of chance, unpredictability and uncontrollability. In light of these chance encounters, canalization and plasticity come into play: we can then conceive of organisms as balancing between maintaining their form and adapting to circumstances.

Many bioethical discussions of epigenetics have explored the relevance of epigenetics to the question of responsibility. For example, what are we to make of the fact that men's behaviour can influence their future offspring's health, long before conception? What does it mean, normatively, that a woman's smoking may affect the respiratory health of her grandchildren? Should a pregnant policewoman quit the stressful job that she enjoys because it has been shown that stress during pregnancy may increase the chance that her child will be hyperactive (Ronald et al, 2010; Dupras and Ravitsky, 2016; Hens, 2017)? These are essential questions, and other contributors to this volume have eloquently written on this topic (see Chapters 2 and 5). However, a shift from genes to environment, from genetic determinism to epigenetics, does not automatically offer an escape from a view of life that tries to reduce phenomena such as behaviours or diseases back to simple explanations. The idea that we are not only defined by our circumstances and lifestyle and by the genes that we inherit from our parents, but also by the experiences of our grandparents and perhaps further generations back in time may lead to an even more reductionist and deterministic view on life. For example, some studies suggest that the trauma of enslaved people lingers in the epigenetic marks of their descendants (Grossi, 2020).

Rather than genetic determinism, we can now talk of epigenetic determinism. Furthermore, there is another danger here. It has been suggested that with the rise of genetic knowledge and since the discovery of CRISPR/Cas9 (a gene editing technology) in 2012, the possibility to select and alter the genes of *in vitro* embryos will open the way to designer babies and eugenics. However, so far, neither genetic knowledge nor genetic technology is currently sophisticated enough to accomplish this. Understanding the mechanisms that govern gene expression may also lead to use of technologies such as CRISPR/Cas9 to amend the epigenetic layer, such as methylation patterns (Nakamura et al, 2021). Changing this epigenetic layer may be technically easier than changing the underlying genes.

In this sense, the idea that the epigenetic layer is malleable may open a more straightforward way into what Juengst and colleagues call epi-eugenics (Juengst et al, 2014). Moreover, as explained in the Introduction to this volume, the related discipline of developmental origins of health and disease (DOHaD) investigates and stresses the importance of environmental influences at conception and *in utero*. The popular media has already reported on some of these findings, such as the claim that eating bacon and eggs during pregnancy will make your child smarter (Mehedint et al, 2010). Brain plasticity research often stresses the importance of the first three years of human life in which everything happens. After that, the window of opportunity closes (Wastell and White, 2017). I am not questioning these findings. The first three years of a child's life are indeed important for further development, and policy makers should ensure that parents and other caregivers have sufficient support to provide the best circumstances for their children. In that respect, such research is informative. However, it would be wrong to think that, if children grow up in sub-optimal circumstances, they are doomed for life, or, conversely, if you get the first three years right, everything will fall inexorably into place. I suggest that Waddingtonian epigenetics urges us to consider the idea that the course of life is unpredictable and offers obstacles and chances at any age. Development is not something that solely happens *in utero* or during the first three years; instead, canalization and plasticity play a role from birth to death. Development and life itself are based on experiences and chance as much as on genes or direct environmental influences.

In the next section, I use the example of autism research to show what such a developmental approach to life can mean for research and ethics in this field and beyond.

Autism research: putting the development back into developmental disorders

I choose autism research to illustrate what the adoption of a developmental and complex systems view on life can mean, for several reasons. First, autism is classified as a 'developmental condition', which is good reason for using it as a starting point for reflection on developmental approaches. Second, for a long time, autism research has focused on genetic causes of autism, and ethical questions have often focused on the desirability of using genetic screening for autism in reproductive medicine (Hens et al, 2016). Third, this focus on causes and explanation has neglected the lived experiences of autistic people.

Research has primarily looked at autism as a developmental disorder in a genetic sense, as something that is caused or at least originates in the prenatal phase. However, if we take the lived experience of autistic people seriously, this means that we should look at autism through the prism of Waddingtonian epigenetics: autistic people have unique life trajectories, and their support needs and wellbeing are highly dependent on the environment in which they live. Researching these dynamics is not only interesting from a scientific point of view, but also goes hand in hand with acknowledging that responsibilities towards autistic people should be assessed on a case-by-case basis and defined by and with autistic people themselves. Hence, an ethics of autism research should not primarily focus on questions of cure, prevention or what to do with genetic knowledge, but rather ensure that what is researched conveys the complexities and situatedness of the autistic experience and is relevant to the flourishing of autistic people.

Autism is classified as a 'developmental disorder' or developmental disability in the American Psychiatric Association's Diagnostic and Statistical Manual of Mental Disorders (DSM-5) (American Psychiatric Association, 2013). Many autistic people object to their condition being called a 'disorder' (Kapp et al, 2013). Hence, I shall not use that term to refer to it further in this chapter. At the same time, it is vital that we take a closer look at what is meant by 'development'. In the diagnostic manual, there is not much explanation about what developmental disorders are. For example, the website of the US Centers for Disease Control and Prevention [1] states that 'Developmental disabilities are a group of conditions due to an impairment in physical, learning, language, or behavior areas. These conditions begin during the developmental period, may impact day-to-day functioning, and usually last throughout a person's lifetime'. Developmental disabilities include autism, attention-deficit hyperactivity disorder, Tourette's and developmental coordination disorder (dyspraxia). However, it is unclear from this definition what is meant by 'the developmental period', and thus the definition is open to a variety of interpretations. It suggests that the impairment at least starts in this developmental period, which is taken to be early in childhood, although that is not specified. However, it is also unclear whether the impairment is caused by something that happens in this period or whether it just becomes apparent in that period. This distinction is not trivial, as is apparent from the history of autism research. During the 1950s and 1960s, it was sometimes argued that autism was caused' by the distant behaviour of parents, especially mothers, and thus a logical 'treatment' would be to remove autistic children from their families and institutionalize them (Nadesan, 2005). Such an approach was harmful to those families for obvious reasons.

Partly as a reaction to the harmful mother-blaming discourse of the 1950s and 1960s, autism has been considered to be a childhood affliction with a primarily genetic cause from the last decades of the 20th century onwards. More recent autism research has hence focused on finding the genes that cause autism. However, in genetic research in general, there has been a shift away from a search for simple genetic causes for traits, dispositions or behaviours. Nowadays, the search for a genetic 'cause' for autism has led to the acknowledgement that 'causes' of autism are complex and cannot be attributed to genes alone (Waterhouse, 2013). Genetic and biological explanations can often help autistic people and their parents accept the diagnosis as an intrinsic part of who they are. At the same time, autistic people have expressed worry that there is an agenda of eugenics behind the promises of genetic research into autism. They fear that the discovery of 'autism genes' could lead to the inclusion of these genes in panels for non-invasive prenatal tests (Sanderson, 2021). Nevertheless, the suggestion that developmental phenomena such as autism may also have an environmental component is often not welcomed by autistic people and their parents. This reluctance is probably due to the lingering ghost of the 'refrigerator mother discourse' (Nadesan, 2005), and the view that many hold that, if something has an environmental component, it can easily be cured, or the person in question is somehow able to change their behaviour at will. An approach to autism inspired by the post-genomics and epigenetics era in biology may help resolve this conundrum.

The fact that autism is understood as a 'developmental disorder' with a strongly genetic component has been taken to mean that the first 'symptoms' arise early in childhood, as this is a criterion in DSM-5 (American Psychiatric Association, 2013). In another diagnostics manual, ICD-10 (World Health Organization, 2004), autism is defined as 'lifelong'. However, what is meant by this innateness and lifelong nature remains unclear. For example, does this mean that autistic people will satisfy the diagnostic criteria throughout their lives? Or is a particular way of being, such as a specific manner of information processing, always there in a person, like a character trait? Such ambiguity is apparent in at least two respects. First, autism research currently buys into the idea of the critical window of the first three years by investigating methods to detect autism early on, even pre-symptomatically, and by investigating what kind of early interventions would work. However, there is much uncertainty about what such interventions would accomplish. Is the aim to make people less autistic, an aim that is challenged by autistic people themselves, or to make them behave in a more typical way? Or should these interventions address the actual challenges autistic people face during their lives?

Second, as is apparent from testimonials of autistic people and their parents, such challenges may differ depending on one's life stage or specific context (Hens and Langenberg, 2018). The life stories of autistic adults demonstrate that what autism means and how it is experienced differ. The factors that contribute to challenges or flourishing depend on the path taken through life. Hence, autism is a truly developmental phenomenon in a Waddingtonian sense: there may be a behaviour or a specific way of being that is canalized and persistent, but, at the same time, it is flexible and dependent on changing circumstances. This implies that research investigating autism as a developmental phenomenon should extend beyond the first three years of life. Waddington's epigenetic landscape and the discussion on epigenesis versus preformation can help to make sense of a developmental approach to developmental disability. In the same way that epigenetics has challenged the 20th-century idea of the gene as a blueprint and has firmly anchored biology as being entangled with the environment, the idea of autism as developmental in a Waddingtonian sense could challenge the research focus on causes and explanatory models to one on interactions and experiences. Such an approach to autism research, querying experiences of stability and change in interaction with what one encounters over the course of a lifetime, could yield a more complete view of the phenomenon of autism but is largely still missing.

A developmental approach to biomedical research does not imply that genetic research has now become obsolete. The move toward a post-genomic and systems biology approach to health needs to factor in life's dynamics to understand the dynamics of genes. At the same time, understanding experiences and life stories can tie in with biological research. In order to understand life, and autistic life in this specific case, biology, environment and experience should be studied not as separate fields but as necessarily entangled. Unfortunately, even though autism researchers acknowledge this context relativity and the dynamics of how autistic people experience their diagnosis, and the importance of studying the dynamics of gene expression in addition to the informational value of genes, there is not much research that incorporates these aspects.

Longitudinal studies investigating the meaning and experience of autism throughout a lifetime are still relatively scarce. A developmental approach also suggests that questions regarding the support of autistic people should be posed differently. Indeed, biomedical science in general should not be seen as separate from societal values; it can influence them, and vice versa. A developmental approach to autism in society means recognizing needs, duties, responsibilities and obligations on a case-by-case basis. It also means adopting a fundamentally inclusive approach to autism management.

It may seem as though I have been attacking a straw man up to this point. Geneticists or other scientists often contend that 'everybody knows' that genetic determinism and reductionism are misguided and that there are environmental components to development. They argue that (post) genetic determinism is merely the result of poor scientific communication in the popular media or a lack of understanding of biology by the general public. My point is not that individual scientists have such deterministic and reductionist views on life but that they may be forced to conduct research in a reductive way.

Let us take it for granted that an important aim of clinical and psychological research is to help the people under investigation. The primary outcome of the search for the 'autism gene' is that there is no such gene but that many genes may play different roles, but the immediate benefit of such knowledge to those studied is uncertain. Many projects now investigate whether autism can be detected earlier using biomarkers or early behavioural observation. However, there is still uncertainty about what to do with this knowledge and how it can help autistic people or even help understand 'autism' as a subject of scientific inquiry. It is increasingly acknowledged that research that aims to 'cure' autism is not what is needed. At the same time, what could help autistic people and how this help may differ between life phases is only recently being incorporated into research. To be clear, my point is not that individual autism researchers or research groups are to blame for the fact that autism research is only now moving away from a purely biological approach . Rather, autism research is an excellent example of how scientific practice and funding today are not ready for a rich understanding of life and experience in all its dynamics and development. Many science projects aim for quick results in a short timeframe, often the time it takes to do a PhD. In order to be successful, project proposals need to be very clear about their end goal and how to achieve it. Finding a gene that is shared in some families with a particular phenotype is feasible in this context, as is identifying general trends in a large population using the statistical methods of behaviour genetics. Spending decades mapping experiences and biology in real life is not so feasible.

A genuinely developmental investigation of developmental disability in all its aspects throughout a lifetime would require a completely different approach to scientific research. It would mean that funding and personnel would have to be guaranteed for many years. Such research could use AI and molecular techniques from systems biology to study the dynamics of gene expression over time, in combination with methods suitable for querying the experiences of the research participants, and with an active engagement of the population under study to co-create research goals and provide feedback during execution of the research. Moreover, such truly developmental science would necessarily be interdisciplinary. In the final part of this chapter, I provide further details on how this approach may shed new light on the role of bioethicists in medical research.

Responsible bioethics, responsible science

In the previous sections, I have suggested that a truly developmental approach to autism in research goes beyond simple causal explanations and instead engages with experiences throughout a lifetime. This means taking lived experiences seriously and looking beyond disciplinary boundaries. In what follows, I suggest what role a bioethicist can play. In short, I argue that bioethics should not involve solely thinking about science, but with science, in a broad endeavour to understand life in all its complexity.

Bioethics as a field is concerned to a large extent with the ethical aspects of biomedical practice. Recently, some bioethicists have critiqued what they call 'mainstream bioethics'. For example, Henk ten Have has argued that bioethics, as it is practised now, has taken too much for granted the science that it is supposed to reflect upon, and has done so without critically reflecting upon the values that underlie scientific and everyday practice (ten Have, 2022). Similarly, it has been argued that philosophical reflection on the concepts and presuppositions of research should be part and parcel of research ethics, and that philosophy of science and bioethics should not be seen as separate endeavours (De Block et al, 2022).

Narrative and care ethics approaches in bioethics have stressed the importance of experiences and the acknowledgement of situatedness of science, practice and ethics (Lindemann et al, 2008). The example of autism research demonstrates that the kind of research that is done and the ontological commitments that it has have direct implications for the good that such research can do. Based on recent and not-so-recent findings in epigenetics and systems biology, I advocate a developmental outlook on life. Such a developmental outlook sees organisms as constantly in flux, and acknowledges the relevance of each life stage and the context in which that life stage takes place. The rise of projects in systems biology appears to corroborate this approach. At the same time, due to current funding practices in science, researchers cannot fully embrace such systemic, longitudinal and developmental approaches. As a companion to science, bioethics follows suit: many bioethics projects and questions focus on specific moments in life (birth, death, illness). Here I make some suggestions for more developmental, systems biology-ready bioethics practice .

What is the responsibility of bioethicists in research projects? From my experience, many research projects in biomedicine assume that the role of an ethicist is, in the first place, a procedural one: their responsibility is to ensure that correct ethical procedures are followed. At the same time, bioethicists can make recommendations on the ethical use of the scientific findings resulting from the research. In this role, they engage stakeholders and investigate opinions. While all these activities are worthwhile, I believe the responsibility of a bioethicist can go beyond this, and that bioethicists may play a role in many aspects of research beyond research ethics.

First, bioethicists have a responsibility to look critically at the concepts taken for granted in research projects. The complex entanglements of biology, experience, culture and society also call for a truly interdisciplinary approach whereby exact sciences, social sciences and humanities work together to make sense of life trajectories. In interdisciplinary projects, bioethicists and philosophers of science may play a role in clarifying concepts and enabling communication between fields. For example, terms such as 'gene' or 'resilience' may have a different meaning if you are a molecular biologist, a behaviour geneticist or a disability scholar. None of these meanings contains the final truth, but philosophers can help build communication bridges. Specifically, I see a role as a 'benevolent gadfly' for the ethicist in a research project.

Gadflies have a bad reputation, both in philosophy and in everyday life. They cause itches. They break our concentration. Bioethicists and philosophers of science can ask 'itchy' questions about the concepts used in research projects, such as the possibility that genes 'predict' autism. Specifically, when it comes to genomics research, bioethicists can uncover assumptions about genes and behaviour in research projects that are simplistic or even reductionist, even when they are not made explicit. For example, if we are engaged in a research protocol that claims to find genetic biomarkers for autism, we can keep questioning what is meant by autism in this case and the link between genes and autism. Does autism refer to a specific neurotype, a behaviour, or challenges that may or may not be associated with a neurotype, depending on the circumstances? In the light of epigenetic findings that challenge simple interpretations of the central dogma, in what way do genes 'cause' behaviour, and what does 'a gene for autism' actually mean?

Second, bioethicists have a responsibility to think critically about how knowledge is acquired. Epigenetics is the molecular proof that one's life course and experiences are entangled with biology. The philosopher Georges Canguilhem said that 'life is experience' (Canguilhem, 2008). Hence, understanding aspects of life that involve people means going beyond searching for explanatory genes. Understanding experiences of wellbeing, resilience and challenges means actively engaging with people who have these experiences. Such co-creation with stakeholders may be challenging and is not always welcomed by researchers. Bioethicists should ensure that stakeholder engagement goes beyond tokenism, and that the voices of those who are often not heard are included. As one reviewer of this chapter rightly stated, many autistic people do not use verbal language, which may impede research participation. However, this does not imply that engagement is a dead end from the start. It is the responsibility of the bioethicist, who is, after all, responsible for the ethics part of the research, to ensure that we try harder to engage those who are hard to engage.

Third, bioethicists have responsibilities that go beyond scientists and research participants. Scientific practice, specifically in the biomedical field, aims (or should aim) to benefit humankind and life at large. The responsibility of ethicists is therefore also to think about the aims and impact of science and the benefits for stakeholders and society. Hence, bioethicists should scrutinize research proposals and not be afraid to ask *cui bono* (who benefits) (Haraway, 1997). For example, what benefit does genetic research on autism offer those being researched? The question *cui bono* is not meant to provoke. It should be asked of both the scientists and the relevant stakeholders to ensure that everyone is on the same page right from the project's inception. Facilitating this communication and co-creation of research aims with those affected by the research should be considered part and parcel of an ethicist's work.

The bioethics gadfly is a benevolent one. It is a friend of science. Rather than stalling or impeding, the benevolent gadfly thinks *with* the scientist. In the end, we may hope that both bioethicists and scientists have the same goal: challenging an old-fashioned reductionist and deterministic view on life and ultimately making science 'better' in many ways for relevant stakeholders.

In this chapter, I have suggested how taking the concept of development seriously and engaging with the complexity of organisms offers a different perspective for bioethics. I want to end with two observations for the reader to ponder. First, the distinction between a preformationist and epigenetic approach, as for related ideas such as nature versus nurture, innate versus acquired, and so on, may be a specifically Western one. It may very well be that certain native American relational ontologies can help us grasp the implications of complexity theory and developmental approaches to life (Cordova, 2003; Cordova, 2007; Gare, 2017). The direction I propose should engage with these ontologies and be inspired by them, otherwise we risk reinventing the wheel. Second, as hinted at in the example of autism genetics, a Waddingtonian approach to (epi)genetics may challenge more than Western preformationist assumptions in bioethics. In fact, it may challenge the very premises of ethical thinking in Western philosophy, such as harm or causality. Given how these concepts are usually linked with common moral standards of responsibility, this should give us pause.

Note
[1] https://www.cdc.gov/ncbddd/developmentaldisabilities/facts.html

Acknowledgements

I would like to thank an anonymous reviewer and Luca Chiapperino for their feedback on earlier versions of this chapter.

References

American Psychiatric Association (2013) *Diagnostic and Statistical Manual of Mental Disorders (DSM-5)*, Washington DC: American Psychiatric Association Publishing.

Canguilhem, G. (2008) *Knowledge of Life*, New York: Fordham University Press.

Cordova, V.F. (2003) 'Ethics: the we and the I', in A. Waters (ed), *American Indian Thought. Philosophical Essays*, Hoboken, NJ: Wiley-Blackwell, pp 173–81.

Cordova, V.F. (2007) *How it is: The Native American Philosophy of V.F. Cordova*, Tucszon, AZ: University of Arizona Press.

De Block, A., Delaere, P. and Hens, K. (2022) 'Philosophy of science can prevent manslaughter', *Journal of Bioethical Inquiry*, 19: 537–43. https://doi.org/10.1007/s11673-022-10198-4.

Dupras, C. and Ravitsky, V. (2016) 'The ambiguous nature of epigenetic responsibility', *Journal of Medical Ethics*, 42(8): 534–41.

Gare, A. (2017) 'Chreods, homeorhesis and biofields: finding the right path for science through Daoism', *Progress in Biophysics and Molecular Biology*, 131: 61–91.

Grossi, É. (2020) 'New avenues in epigenetic research about race: online activism around reparations for slavery in the United States', *Social Sciences Information*, 59(1): 93–116.

Haraway, D.J. (1997) *Modest_Witness@Second_Millennium.FemaleMan_Meets_OncoMouse: Feminism and Technoscience*, New York: Routledge.

Hens, K. (2017) 'The ethics of postponed fatherhood', *International Journal of Feminist Approaches to Bioethics*, 10(1): 103–18.

Hens, K. and Langenberg, R. (2018) *Experiences of Adults Following an Autism Diagnosis*, Berlin: Springer.

Hens, K., Peeters, H. and Dierickx, K. (2016) 'The ethics of complexity. genetics and autism, a literature review', *American Journal of Medical Genetics*, 171(3): 305–16. https://doi.org/10.1002/ajmg.b.32432.

Jablonka, E. (2016) 'Cultural epigenetics', *The Sociological Review*, 64(suppl 1): 42–60.

Jablonka, E. and Lamb, M.J. (2014) *Evolution in Four Dimensions, Revised Edition: Genetic, Epigenetic, Behavioral, and Symbolic Variation in the History of Life*, Cambridge, MA: MIT Press.

Juengst, E.T., Fishman, J.R., McGowan, M.L. and Settersten, Jr, R.A. (2014) 'Serving epigenetics before its time', *Trends in Genetics*, 30(10): 427–9.

Kapp, S.K., Gillespie-Lynch, K., Sherman, L.E. and Hutman, T. (2013) 'Deficit, difference, or both? Autism and neurodiversity', *Developmental Psychology*, 49(1): 59–71.

Lindemann, H., Verkerk, M. and Urban Walker, M. (2008) *Naturalized Bioethics: Toward Responsible Knowing and Practice,* Cambridge: Cambridge University Press.

Mehedint, M.G., Niculescu, M.D., Craciunescu, C.N. and Zeisel, S.H. (2010) 'Choline deficiency alters global histone methylation and epigenetic marking at the Re1 site of the calbindin 1 gene', *The FASEB Journal*, 24(1): 184. DOI: 10.1096/fj.09-140145.

Nadesan, M.H. (2005) *Constructing Autism: Unravelling the 'Truth' and Understanding the Social*, London: Routledge.

Nakamura, M., Gao, Y., Dominguez, A.A. and Qi, L.S. (2021) 'CRISPR technologies for precise epigenome editing', *Nature Cell Biology*, 23(1): 11–22.

Parfit, D. (1984) *Reasons and Persons*, Oxford: Oxford University Press.

Ronald, A., Pennell, C.E. and Whitehouse, A.J.O. (2010) 'Prenatal maternal stress associated with ADHD and autistic traits in early childhood', *Frontiers in Psychology*, 1: 223.

Sanderson, K. (2021) 'High-profile autism genetics project paused amid backlash', *Nature*, 598(7879): 17–8. https://doi.org/10.1038/d41 586-021-02602-7.

Tavory, I., Jablonka, E. and Ginsburg, S. (2014) 'The reproduction of the social: a developmental system theory approach', in L.R. Caporael, J.R. Grisemer and W.C. Wimsatt (eds), *Scaffolding in Evolution, Culture and Cognition: Vienna Series in Theoretical Biology*, Cambridge, MA: MIT Press, pp 307–27.

ten Have, H.A.M.J. (2022) *Bizarre Bioethics: Ghosts, Monsters, and Pilgrims*, Baltimore, MD: Johns Hopkins University Press.

Varela, F.J., Rosch, E. and Thompson, E. (1992) *The Embodied Mind: Cognitive Science and Human Experience*, Cambridge, MA: MIT Press.

Waddington, C.H. (1948) *The Scientific Attitude*, London: Penguin Books.

Waddington, C.H. (1977) *Tools for Thought*, London: Cape.

Waddington, C.H. (2012) 'The epigenotype. 1942', *International Journal of Epidemiology*, 41(1): 10–13.

Wastell, D. and White, S. (2017) *Blinded by Science: The Social Implications of Epigenetics and Neuroscience*, Bristol: Policy Press.

Waterhouse, L. (2013) *Rethinking Autism: Variation and Complexity*, Cambridge, MA: Academic Press.

Wessel, A. (2009) 'What is epigenesis? Or Gene's place in development', *human_ontogenetics*, 3(2): 35–37.

World Health Organization (2004) *ICD-10: International Statistical Classification of Diseases and Related Health Problems – tenth revision* (2nd edn), World Health Organization. Available from: https://apps.who.int/iris/handle/10665/42980 [Accessed 3 May 2023].

2

Epigenetics and Forward-Looking Collective Responsibility

Emma Moormann

Introduction

This chapter is concerned with the ethics of epigenetics from an egalitarian perspective. Our societies are currently deeply unequal in the ways in which resources, opportunities and exposure to harmful phenomena are distributed. Disparities and injustices are also present in the occurrence and distribution of epigenetically mediated harm. One does not have to be an epigenetic exceptionalist (Rothstein, 2013; Huang and King, 2018) to contend that findings in epigenetics are another addition to the vast wealth of empirical evidence showing that social inequalities have an impact on individuals and their offspring, both physically and mentally.

When thinking through issues of justice with regards to public health in general, and epigenetics in particular, the concept of responsibility has often proven to be an indispensable tool. In this chapter, I aim to add to the literature on epigenetics and responsibility by focusing on a specific group of responsibilities: forward-looking collective responsibilities (FLCR). I explore how the concept of FLCR can contribute to a balanced account of responsibility in the context of epigenetics.

As Smiley explains, what is specific about forward-looking responsibilities is that they are 'ascribed for the purpose of ensuring the success of a particular moral project rather than for the purpose of gauging the moral agency of a particular group' (Smiley, 2014, p 6). Such an approach does not focus on the question of who has caused a current state of affairs. Rather, it aims to find suitable individual or collective agents who can take responsibility for bringing about a desirable state of affairs.[1]

Many long-lasting debates about collective responsibility are ongoing, focusing on two questions. First, do truly collective agents exist? Second, is it fair, useful or appropriate to ascribe responsibility to collective agents? Methodological and normative individualists give a negative answer to both questions. However, in this chapter, I focus on those authors who do think it is a feasible and appropriate concept to use in normative philosophical debates. I acknowledge the debate and realize that the premise of this chapter, namely that collective agents can be bearers of responsibility, is not self-evident. However, entering into the debate about whether there can even be such a thing as collective responsibility is beyond the scope of the chapter. Thus, I presuppose that collective agents can be the legitimate bearers of responsibility in my argument in favour of the usefulness of FLCR in thinking about the ethics of epigenetics.[2]

While I do presume that collective responsibility exists, and hence do not argue for this point, I later present a defense of the reasonableness of FLCR in ethical questions concerning epigenetics. The concept of FLCR has quickly gained ground as a tool to discuss new complex global problems, as well as generation-superseding problems, such as racism and climate change. The health impact of epigenetic mechanisms is an equally complex phenomenon, analysis of which would benefit from the concept of FLCR. Moreover, although it is relatively absent in discussions on the ethics of epigenetics, FLCR has already been identified as relevant and promising by a number of authors in the context of epigenetics (Hedlund, 2012; Dupras and Ravitsky, 2016; Chiapperino, 2018; Meloni and Müller, 2018).

The next section of this chapter provides an overview of existing debates about collective responsibility, and specifically FLCR, in the literature on the ethics of epigenetics. I develop my own set of recommendations for using FLCR in this context. To an important degree, I reach these recommendations by means of applying insights from more general philosophical accounts in political philosophy and analytic responsibility theory to the specific challenges that arise from epigenetic knowledge. I argue for the following claims:

- We need to steer clear of epigenetic eliminativism – the idea that in light of epigenetic findings, we should refrain from any responsibility ascriptions;
- FLCR is particularly well-suited to an ethical account that strives towards epigenetic justice; conversely, epigenetic injustice may fruitfully be understood as an instance of historical–structural injustice;
- intersectional feminist thinking, and particularly disability justice work, provides useful tools for the analysis of epigenetic justice;
- FLCR ascriptions may be based on a variety of sources and concerns. FLCR is only useful when integrated into an account of epigenetic responsibility that also leaves room for backward-looking concerns.

The final part of this chapter illustrates these recommendations in a real-life context by discussing some epigenetic mechanisms in action in Mexico City.

A few notes on terminology are in order. First, unless otherwise stated, when I talk about responsibility, I have in mind moral responsibility rather than, for example, legal or causal responsibility, although all these concepts may be intricately connected when it comes to complex structural problems such as epigenetic (in)justice. Second, although not all scholars referred to in this chapter use the term 'forward-looking (collective) responsibility', their accounts are nonetheless within the purview of this analysis because they meet two requirements, that (1) they are in some sense forward-looking (primarily concerned with future states of affairs), and (2) they allow responsibility to be ascribed to a collective agent.

FLCR and the ethics of epigenetics

This chapter assumes that ethical epigenetic exceptionalism (see Introduction to this volume) is unwarranted. The ethics of epigenetics are not so fundamentally different from those of other complex bioethical or public health issues in that they require the use of separate concepts of responsibility. Moreover, some of the characteristics of epigenetics warrant a search for concepts of responsibility that are being used in normative work on global issues such as climate change and structural racism. Although they are not specific to epigenetics, especially not when taken individually, four characteristics of epigenetics are particularly relevant for the development of a responsibility framework. These characteristics are: (1) the role of the environment (broadly understood) in the health of an organism at the molecular level, (2) the possibility, although still contested, of transgenerational inheritance, (3) causal complexities and uncertainties that make it very hard to define epigenetic harm or health, and (4) the potential reversibility of epigenetic mechanisms.

There has been a lively debate with regard to which kind of responsibility concepts to use when discussing the ethically salient characteristics of epigenetics. An emphasis on individual responsibility is often criticized because it is believed to be unfair in light of the complex connection between individual choices and changes to the epigenome (Hedlund, 2012; Heijmans and Mill, 2012; Mill and Heijmans, 2013). Dupras and Ravitsky share the concern that 'some scholars, the public and the media are at risk of too hastily and simplistically assigning most epigenetic responsibilities to individuals' (Dupras and Ravitsky, 2016, p 6). However, they are equally wary of simplistic prospective and state-focused solutions. Instead, they propose a 'diversity of types' of epigenetic responsibility that can deal with the nuances regarding the definition of a 'normal' or healthy reference epigenome in a specific context (epigenetic normality) and the dynamic nature of epigenetic modifications (epigenetic plasticity).

Chiapperino provides another version of this critique based on the influence of moral luck on individual agency. Moral luck consists of 'the import that factors beyond one's control have on the justification and cogency of normative claims such as responsibilities' (Chiapperino, 2020, p 2). He goes a step further, however, by showing that much of the critique of individual responsibility in the context of epigenetics also applies to collectives. He argues that it may be unwarranted to exempt collectives 'from a consideration of how intrinsic limitations and deficiencies, trying and unwanted circumstances, as well as imperfectly predictable results temper their blameworthiness for failing to act responsibly to protect our epigenomes and health' (Chiapperino, 2020, p 12; see also Chapter 3). For instance, it may be hard to determine the contributory liability or backward-looking responsibility of individual members of a collective. Furthermore, it is often unclear to what extent past and present members of a collective have contributed to its actions leading to certain epigenetic effects. This concern is not unique to epigenetics. It has also been raised, for example, to criticize calls for 'corrective justice' in dealing with climate change (Posner and Sunstein, 2007).

The appropriateness of forward-looking collective responsibility specifically has been discussed by various authors interested in the ethics of epigenetics. The first substantive account of FLCR in an epigenetics context is that of Maria Hedlund. She argues that epigenetic responsibility should primarily be collective instead of individual (Hedlund, 2012). She then draws on the social connection model of responsibility proposed by Young (discussed later in this chapter) to argue for prospective political responsibility to be ascribed primarily to state institutions. Whereas backward-looking models strive to isolate a responsible agent, a forward-looking model 'will tend to disregard the structural factors that shape the norms of appropriate behaviour and that an integrated forward-looking responsibility model brings into question' (Hedlund, 2012, p 179). According to Hedlund, 'the moral dimension of solidarity justifies why agents with capacity or in a position to act should be responsible in a forward-looking way' (Hedlund, 2012, p 178). If we care about equality, and if we value solidarity with the worse off, we should pay more attention to forward-looking collective epigenetic responsibilities.

Dupras and Ravitsky (2016) are critical of FLCR approaches. They are sceptical not only of the focus on collective instead of individual agents, as mentioned above before, but also of any account that is exclusively forward-looking. Such an account would be ineffective if put into practice, because 'attributing mere prospective responsibility without the possibility of holding actors responsible for past negligence (through health policies or laws) may result in a very limited upholding of the suggested prospective responsibility'

(Dupras and Ravitsky, 2016, p 3). Put simply: if there are no consequences for not doing what is prospectively required, neither individual agents nor institutions will be very motivated to act on their responsibility by making efforts that may well be costly for them in some way (Neuhäuser, 2014).

Chiapperino (2020) also discusses forward-looking collective responsibilities. He argues that the criticism that can be levelled against the notion of collective responsibility in general also applies to the use of specific forward-looking collective responsibilities, albeit in a somewhat different form. First, he criticizes accounts that equate 'remedial collective responsibility to the capacity [to take] informed action about a given situation' (Chiapperino, 2020, p 10). Second, he reminds us that 'appeals to forward-looking collective responsibilities do not automatically support the idea that action should tackle the structural configurations of society producing epigenetic hazards' (Chiapperino, 2020, p 10). Perhaps other solutions or approaches, such as personalized medical interventions on an individual level, might be more appropriate. Third, he argues that collective agents are exposed to 'contingencies and circumstances of agency or the stochastic and highly contextual dependency of epigenetic predispositions to disease' (Chiapperino, 2020, p 10) no less than individuals are. The outcome of the actions of collective agents may be influenced by factors outside policy control, just as the outcomes of individual actions may be influenced by structural factors.

Suggestions towards a framework

Against (epigenetic) eliminativism

If responsibility concepts in the context of epigenetics are so fraught with problems, is developing recommendations for a framework of epigenetic responsibility futile? Concerns such as those summarized above, directed against both individual and collective responsibility, may tempt those thinking through the ethical implications of epigenetics to become 'epigenetic eliminativists' with regard to responsibility.[3] That is to say, it may be tempting to conclude that epigenetics is simply too complex to factor into the usual mechanisms for ascribing responsibility, which is why we may need to refrain from ascribing responsibility in the context of epigenetics altogether.

I do not endorse epigenetic eliminativism, and agree with Mich Ciurria that we cannot really do without our responsibility practices (Ciurria, 2019). The best we can do in the face of existing flaws in responsibility models is to correct them in a reasonable way. Inspired by her commitment to intersectional feminism, Ciurria argues for a radical transformation of the responsibility system rather than its eradication in the face of the problems embedded in our current responsibility practices. She states her position as follows: 'Whereas eliminativism seeks to address the problem of excessively punitive blame, intersectional feminism instead identifies the core problem as a matter of asymmetrical power relations' (Ciurria, 2019, p 227). Importantly, in such asymmetrical circumstances, responsibility practices can contribute to the emancipation of those people or groups who are holding others responsible for something.

Even Chiapperino, who criticizes several accounts of epigenetic responsibility in his 2020 paper and in his contribution to this volume (see Chapter 3), does not seem to want to opt for an epigenetic eliminativism of responsibility. He argues that 'dominant atomistic framings' (Chiapperino, 2020, p 13), in which either individual or collective agents are central, fail to do justice to the entangled reality of our lives, bodies and environments. But instead of moving away from responsibility ascriptions altogether, he emphasizes 'the need [for] delineating pragmatic, conventional or role collective responsibilities, based on distributive theories of agency, on accessory justifications of autonomy, solidarity, vulnerability and human flourishing, or other norms of our moral and political life' (Chiapperino, 2020, p 13) that could guide collectives in taking up their responsibility.

We still need to work with responsibility concepts, imperfect as they may be. I suspect that leaving some room for responsibility ascription and distribution may be more effective than, for example, arguing for increased unspecified solidarity with regard to public health. Although this may be a noble endeavour, perhaps few agents would be inclined to take action on the basis of such a call for solidarity. This could lead to a kind of action void that undermines their motivation to act. Moreover, an emphasis on individual responsibility for health prevails in public debates as well as many scholarly discussions. I cherish the hope that sufficient attention to collective responsibility in the context of epigenetics will help to provide some counterweight to this focus on individual responsibility.

Epigenetic justice

If we accept that responsibility concepts have a place in the ethical analysis of epigenetics, can FLCR also play a role? If so, what could we hold agents responsible for? As the nature of FLCR is primarily prospective or forward-looking, its objective needs to be some desirable future state of affairs. Because of its collective nature, its aim need not be identified on an individual scale. Instead, it may be a goal to be aimed for at a societal level. I propose that a suitable object of epigenetic FLCR is targeting epigenetic injustice and striving towards epigenetic justice.

This claim is compatible with other claims regarding epigenetic responsibility. First, it is compatible with the idea that a framework for epigenetic responsibility should also encompass backward-looking concerns. Second, it does not deny the validity of pursuing other responsibility objects. Other kinds of epigenetic responsibility objects may in fact be better served by other responsibility concepts. For example, striving towards epigenetic justice can accompany the search for cures for epigenetically mediated diseases. The responsibility claims pertaining to such cures may well be situated on a more individual level, such as that of personalized medical interventions. Other targets for epigenetic responsibility that may exist alongside (or be sometimes in tension with) a collective focus on epigenetic justice include responsibilities to prevent adverse epigenetic alterations, to avoid epigenetic harm, or to protect one's own epigenome or that of one's offspring.

Epigenetic injustice may be characterized as a kind of historical–structural injustice. The political philosopher Iris Marion Young defines a situation of structural injustice as one in which 'some people's options are unfairly constrained and they are threatened with deprivation, while others derive significant benefits' (Young, 2011, p 52). Such situations often arise because individuals and institutions pursue their own goals and interests to the detriment of others 'for the most part within the limits of accepted rules and norms' (Young, 2011, p 52). Although epigenetic changes take place at the molecular level of individual organisms, their environmental causes and the distribution of their occurrence between populations are structural issues. Being responsible in relation to such structural injustice is primarily forward-looking and collective in nature. Each of us has a political responsibility to 'transform the structural processes to make their outcomes less unjust' (Young, 2011, p 96) that can only be discharged through collective action.

More specifically, epigenetic injustices are instances of what Nuti (2019) terms historical–structural injustices, which she defines as 'unjust social-structural processes enabling asymmetries between differently positioned persons, which started in the past and are reproduced in a different fashion, even if the original form of injustice may appear to have ended' (Nuti, 2019, p 44). Epigenetic injustice may be characterized as a biosocial instance of the kind of historical–structural injustice that she discusses. Skewed distributions of ill health may have historical roots. For example, epigenetics research has been used to study the biological basis of intergenerational trauma of indigenous Australians as a result of the harmful effects of actions undertaken by colonial forces (Warin et al, 2020). Epigenetic knowledge about the intricate connections between biography and biology can help us to understand how persistent current health disparities can be. Moreover, Nuti's term 'banal radicality' applies very well to the reproduction of epigenetic injustice. The ways in which injustice is reproduced in the present are banal in the sense that individual acts contributing to them are not easily perceived as clearly objectionable or wrong. Reproductions are often subtle, difficult to point out, and sometimes even unconscious, but those small, often unintended, reproductions together 'provide the condition of possibility for radical injustices to occur' (Nuti, 2019, p 44). This idea can help us to understand how many actions by people and institutions, interactions with other agents, and short- or long-term exposure to factors such as pollutants and toxins, affect people's health by/through epigenetics. Fortunately, the potential reversibility of epigenetic mechanisms provides the promise of actionability to those striving towards epigenetic justice.

Intersectional feminism and disability justice

Epigenetic justice is not yet a substantive goal for ascribing and accepting epigenetic responsibility. Conceptions of justice may differ considerably. This section outlines what applying an intersectional feminist normative perspective might entail in the context of epigenetics.

Feminist theory has its roots in the struggle against oppressive gender roles, misogyny and sexism, but it also addresses the intersections of the power dynamics involved in gender with other forms of subordination (Allen, 2021). Consequently, feminist theory is closely linked to areas such as class analysis, critical race theory, queer theory and (critical) disability theory.

Ciurria identifies five central aims of an intersectional feminist theory that may be used in approaching the topic of moral responsibility (Ciurria, 2019). Such an approach should be aimed at (1) foregrounding and diagnosing the intersection of injustice, oppression and adversity, and (2) actively combating them. To do so, researchers can use (3) an ameliorative method that 'defines concepts partly by reference to normative goals that challenge the status quo' (Ciurria, 2019, p 4).[4] Ciurria also (4) urges us to take up Charles Mills' call to strive towards a non-ideal theory (Mills, 2005), which has the advantage that it 'avoids abstractions that misrepresent reality' (Ciurria, 2019, p 5), unlike ideal theory that assumes just background conditions. Finally, she characterizes an intersectional feminist approach as one that is committed to (5) a relational method. When it comes to complex responsibility issues, we should develop relational explanations that combine 'situational and agential, collective and individual levels of analysis in a holistic fashion, giving rise to an understanding of individual responsibility as a function of the individual's role in situations and collectives' (Ciurria, 2019, p 229).

In the context of epigenetics, we can ask which groups are suffering the most harm as a result of epigenetically mediated influences, and why. How should the benefits of epigenetic knowledge be distributed? It is likely that privileged groups are better placed to benefit from epigenetics research – they may be able to reverse some adverse epigenetic alterations by getting the right nutrition and supplements, living in a favourable environment, and, increasingly, seeking the right treatments. Underprivileged groups, on the other hand, tend to be impacted disproportionally by the environmental triggers that cause adverse health effects through epigenetic mechanisms. For example, epigenetic mechanisms have been implicated in the link between low socio-economic status and poor health status, also known as the 'Glasgow effect' (McGuinness et al, 2012; Katz, 2018; Vears and D'Abramo, 2018). Also, environmentally induced adverse health effects mediated by epigenetic mechanisms affect people of colour disproportionally (Sullivan, 2013; Mansfield and Guthman, 2015; Saulnier and Dupras, 2017). Findings in epigenetics linking the influence of ancestral trauma with current health problems have also been cited by activists demanding reparations for slavery (Grossi, 2020; Warin et al, 2020).

I discuss these issues further in a later section of this chapter, in relation to some examples of epigenetic responsibility. For now, I move to a discussion of the relevance of intersectional theory for epigenetic responsibility, focusing on the intersections between epigenetic justice and disability justice.

Disability theory and epigenetics

Saulnier (2020) argues that insights from disability studies are urgently needed to question some key assumptions in the ELSI (Ethical, Legal and Social Implications) literature on epigenetics. Saulnier notes that 'epigenetics as an emerging field is already showing a tendency to feed into harmful narratives around the value of certain bodies over others' (Saulnier, 2020, p 13). For instance, epigenetics research often seems to take for granted a focus on identifying and discussing epigenetic 'deficits' or 'defects' (Saulnier, 2020, p 25).

Disability studies and critical disability theory offer tools to delineate what counts as epigenetic harm, and to distinguish between harmful, neutral and beneficial epigenetic variation. Such insights are very helpful for those thinking about responsibility for bringing about epigenetic justice. Developing a substantial concept of justice also implies a certain understanding of epigenetic harm, as something to be avoided, mitigated or distributed more fairly. Epigenetic harm often functions as a kind of 'bridging concept' between relatively neutral findings in the field of epigenetics (such as findings about the workings of specific epigenetic mechanisms) on the one hand, and potential ethical and societal implications of those findings, primarily in terms of responsibility ascription and distribution, on the other.

Because it is very difficult to define a healthy or normal epigenome (Dupras et al, 2019; Santaló and Berdasco, 2022), 'we should be careful not to conflate the atypical epigenome with the detrimental' (Dupras and Ravitsky, 2016, p 4). Furthermore, it is extremely difficult to solve the complex puzzle of discovering which elements of a person's lifestyle, environment and genetic make-up contribute to certain epigenetic alterations (Chiapperino, 2018; Chiapperino, 2020).[5] In addition to such epistemological concerns, some disability theorists also raise ethical concerns. They are critical of a medical model of disability that 'frames atypical bodies and minds as deviant, pathological, and defective, best understood and addressed in medical terms' (Kafer, 2013, p 5). In the context of epigenetics, adherence to such a model becomes apparent in the search for cures and therapies for certain conditions, given the promise of reversibility of epigenetic mechanisms. Some disability theorists, such as Elizabeth Barnes, point out that it is wrong in general to assume an implicit connection between bodily differences – epigenetic or otherwise – and deficits. They often hold variations of the 'mere-difference view', arguing that disability is not in itself something that always makes disabled people worse off (Barnes, 2014).

But if not all instances of epigenetic variation count as epigenetic harm, how can we determine which ones do? One key approach to answering this question lies in paying attention to the lived experience of people with conditions to which epigenetic mechanisms have contributed (Hens and Van Goidsenhoven, 2017). Critical disability theory rests on the belief that, in general, people with first-hand experience of the impact of environmental factors and lifestyle behaviours on their bodies are best placed to judge whether they have been harmed. In the context of epigenetics, this implies that one needs to listen to people's accounts of their own quality of life, the ways in which they experience their interaction with their environment, and the obstacles they encounter. Paying attention to lived experience can alter our understanding of epigenetic harm by constraining it in some respects and broadening it in others (for example, the neurodiversity movement rejects the idea that autism or attention-deficit hyperactivity disorder are self-evidently conditions to be prevented).[6]

A disability lens may help us to not only consider the direct impact of epigenetic changes, but also the indirect harms that may result from such changes. For example, Saulnier and colleagues argue convincingly that focusing epigenetic research on already vulnerable or minority groups could result in stigmatization:

> [In epigenetics research], populations that have experienced large scale trauma or early-life adversity are being examined to provide evidence of the patterns already noted by researchers in other medical and social science fields. In providing a new layer of evidence for existing observations of health precarity and reduced health outcomes for populations that face discrimination, stigmatization, and trauma, researchers risk reifying stereotypes and placing contestable normative values on cultural behaviours or cognitive differences. (Saulnier et al, 2022, p 69)

If epigenetics research does not sufficiently respond to the disability justice slogan 'nothing about us, without us', it risks inflicting further harms on people who are already disadvantaged.

Dimensions and sources of collective responsibility

Intersectional feminism and disability perspectives are thus one potential normative lens through which one can study the applicability of FLCR for epigenetic justice. Such a lens may help to adjudicate 'the salience of various practical and normative considerations' (Smiley, 2014, p 11) in a particular case. In this section, I discuss some relevant practical and normative considerations for the ascription of collective responsibility towards epigenetic justice.

Orientation and justification

First, a helpful distinction that allows for a nuanced use of the concept of FLCR is the one made by Linda Radzik between two dimensions of responsibility: orientation and justification (Radzik, 2014). The orientation of responsibility focuses on what the agents pays moral attention to (which concerns lead them to act) or the character of the responses that they make to a certain state of affairs (Radzik, 2014, p 32). It is not enough to rely on established moral rules in a forward-looking orientation: one 'can only fulfil one's responsibility through a more open-ended engagement with the possibilities the future might hold' (Radzik, 2014, p 36).

The justification dimension of responsibility denotes 'the kind of reason or justification that the victim and the community have for responding [to, for example, an action or behaviour] the way we do' (Radzik, 2014, p 33). Backward-justified responsibility claims may be supported by desert-based or justice-based claims, and are less concerned with the positive consequences of ascribing or taking up certain responsibilities. Forward-looking justifications, in contrast, are those that are justified by appealing to consequences or pragmatic considerations (Radzik, 2014, p 34).

In this way, Radzik helps us to understand that, although forward- and backward-looking dimensions are entangled, they are analytically distinguishable. Consider two statements about FLCR concerning the fictional Universal Corporation (UniCorp), which releases excessive amounts of lead particles into the air as a side-effect of its activities, thereby triggering epigenetic mechanisms leading to a higher prevalence of neurological disorders in those living close to the factory. We might think that:

- UniCorp should accept responsibility for avoiding epigenetically induced adverse health effects in future generations;
- or that the local government should pay compensation to those people who grow up in the vicinity of UniCorp, because the government failed to enforce sufficiently stringent measures to prevent the corporation's damaging activities.

Holding an agent responsible for avoiding future harms, as in the first example, is forward-looking in both dimensions. However, the source of an agent's responsibility may also be more backward-looking in nature, such as the complicity of the government in the second example. The object of epigenetic justice gives FLCR a forward-looking orientation. However, the justification of its ascription may be based on both forward- and backward-looking concerns. Moreover, one responsibility ascription may have multiple sources. I now discuss some examples.

Sources of normative responsibility

Björnsson and Brülde define normative responsibilities as the requirement to care about what one is responsible for (Björnsson and Brülde, 2017). Such responsibilities 'are themselves primarily prospective, and are often grounded in what can be done rather than in what has been done' (Björnsson and Brülde, 2017, p 14). They may also be attributed to collective agents, making them useful for our FLCR-focused account. The authors provide a list of six distinct potential sources of normative responsibility. Each of them may be relevant to consider when ascribing FLCR in the context of epigenetics:

- *Capacity and cost.* Responsibility may be ascribed to agents because we believe them to be particularly well placed to take on a task or solve a problem. For example, Dupras and Ravitsky talk about 'windows of opportunity' in the context of epigenetics, arguing that efficient preventive or curative interventions require that 'moral epigenetic responsibilities should be recognized as necessarily context-dependent and relying on who has a capacity to act' (Dupras and Ravitsky, 2016, p 5). At the same time, Chiapperino argues (2020) that, at least in the context of epigenetic FLCR, a narrow focus on this source is insufficient, because the capacity of collectives to bring us closer to epigenetic justice can easily be overestimated.
- *Retrospective and causal responsibility.* Other things being equal, a greater causal backward-looking responsibility (through causal connections) is positively correlated with the degree of forward-looking responsibility. This may even be the case if the agent was merely involved in creating a risk of harm. As Marion Smiley remarks, we need to acknowledge that there will almost always be multiple candidates for causal status with respect to harm (Smiley, 2014). Determining the exact degree of causal contribution is often nearly impossible, especially in complex cases such as racism, poverty, or indeed epigenetic harm.
- *Benefitting.* Benefitting from someone's help may create a responsibility on the part of the beneficiary to return the favour. But benefitting may also take the form of complicity, when agents benefit from harm, injustice or

danger to others. Just as we may want to 'hold corporations responsible for the profits they derived from slavery' (Young, 2011, p 175), we can ascribe FLCR to organizations on the grounds of their having benefitted from environmental pollution.[7]

- *Promises, contracts and agreements.* If an agent has voluntarily agreed to do something, they are in principle responsible for doing it.
- *Laws and norms.* Epigenetic justice can and should be translated from moral into legal responsibility ascriptions if necessary. Paying attention to legal prescriptions as a potential source for moral responsibility ascriptions can address the concern of Dupras and Ravitsky regarding the upholding of prospective responsibility (see previous section).
- *Roles and special relationships.* We may have special responsibilities by virtue of our social or professional roles, for example our roles as parents or our membership of a specific community (see also Chapter 5).[8]

This list has heuristic value; it helps those working on the ethics of epigenetics or other complex global challenges to look for a broad variety of agents to whom (forward-looking collective) responsibility ((FLC)R) can be ascribed. I contend that (FLC)R is most useful when embedded in an integrated approach to epigenetic responsibility that does not rule out the legitimacy of more backward-looking concerns such as 'retrospective and causal responsibility' or 'promises, contracts and agreements'.

Epigenetics in Mexico City

The research of Elizabeth Roberts in Mexico City illustrates how some of the points made in this chapter may be applied to a specific case. Roberts is an ethnographer of science, medicine and technology who collaborates with the 'Early Life Exposure in Mexico to Environmental Toxicants' (ELEMENT) project, in which environmental health researchers are working together with public health officials. Since 1993, the project team have collected numerous samples for molecular analysis (epigenetic and otherwise), primarily from working-class mothers and children (Roberts, 2015a). In one of the most polluted cities on earth, they are looking into the impact of environmental toxins on multiple generations.

In her work, Roberts focuses on the following features of her subjects' lives: the use of lead-glazed plates, consumption of soda, and proximity to a dam filled with waste. Her fellow researchers found that eating off traditional lead-glazed plates, which are said to make the food taste sweeter, was the surest predictor of high lead levels in mothers and children (Téllez-Rojo et al, 2002; Roberts, 2019). The link between lead exposure and epigenetic alterations is well-established (see for example Senut et al, 2012; Wang et al, 2020). The exposure to lead is gendered, because women are the ones cooking with these utensils and inheriting them from their (grand) mothers, as well as cultural, because the plates connect their users to a rural past (Roberts, 2019). Additionally, the high consumption of sweets and sugary soda is said to be an important factor in the high obesity and diabetes rates in the poorer neighbourhoods of Mexico City such as the one being studied (see, for example, Rosen et al, 2018 for associations between epigenomic changes and obesity and (pre)diabetes). Soda is almost as cheap as bottled water, and is more reliably available than tap water (Roberts, 2017). It performs important social roles, because 'in Moctezuma sharing soda, liquid-food, filled with sugar, is love' (Roberts, 2015b, p 248). Finally, there is a strong smell in the neighbourhood, caused by 'a narrow stream of dam runoff, filled with *aguas negras* (untreated sewage) and garbage' (Roberts, 2015b, p 592). In rainy seasons, the dam often overflows, leaving the walls of the cement houses impregnated with salmonella, *Escherichia coli* and faecal enterococcus (Roberts, 2017, p 593). Whether the dam also causes respiratory diseases is hard to say, according to Roberts, because respiratory problems are commonplace in the whole polluted city.

We can regard the ill health of the inhabitants of the neighbourhood Colonia Periférico at least partly as a matter of epigenetic, historical–structural injustice. It is not illegal that soda is so cheap. Although safe tap water is not universally available, access to it is better than in the past. Nonetheless, past injustices continue to leave their mark, as bottled water and soda companies still profit from the belief that tap water is unsafe (Roberts, 2017). An intersectional lens is helpful in understanding how those problems are connected with many other axes of inequality and oppression. Women's socially ascribed roles as housewives make them more vulnerable to the effects of lead exposure. It is also hard to imagine that the terrible pollution in the neighbourhood would still be accepted in a society without such rampant socio–economic inequalities. Intergenerational justice appears relevant too; the fact that epigenetic mechanisms are involved may well mean that, even if the current environmental hazards are successfully minimized, future generations will still bear the biological marks of the hazards to which their parents and grandparents have been exposed.

What can we say about forward-looking collective responsibility in this context? In cases such as this, collective responsibility ascriptions are to be preferred over an eliminativist approach in order to forefront the injustices that shape the situation in Mexico City. Such a focus on collectives provides a counterweight to the individualist, blaming and stigmatizing responsibility discourse used by government campaigns 'exhorting you as an individual, female, *ama de casa* (housewife) to stop heedlessly providing soda and junk food to your child' (Roberts, 2015b, p 247).

A reluctance to impose individual responsibilities should not preclude us from expecting parents to take up forward-looking role responsibilities towards their children's health. This can still be part of an approach based on intersectional feminist concerns. However, we should be very clear about the structural constraints on individual behaviour and choice. It is important that collectives should be urged to take on responsibility, not only for the sake of fairness, but because these are the organizations that most obviously have the potential to make a difference.

Corporate agents and governments can be encouraged to take up responsibility on a variety of grounds. The first is their involvement in bringing about or maintaining current injustices (retrospective/causal responsibility and benefit; Björnsson and Brülde, 2017). Government agencies may also be held responsible for improving the state of the dam water, because they are the guardians of public health. Academics and healthcare providers also have their role to play. For example, epigenetics researchers such as those in the ELEMENT project are using their knowledge and skills to call attention to health disparities.

Additionally, other agents such as anthropologists may use their methodological skills to work together with research participants to tackle difficult problems. Members of the ELEMENT project helped some women to make the decision to switch to metal pots by working together with potters (Roberts, 2015b, p 247). Roberts provides valuable perspectives on the lead-glazed plates issue:

> Participants tell me that the pots were less damaging when the world was less damaged. Their grandparents and great-grandparents made and ate off the pots into their 90s, and were whip smart and not neuro-affected until the end. Now they are forced to reconfigure their relationships to the pots and to each other in light of the fact that there is more contamination all around. They also must grapple with the fact that the pots now have more lead because, with less available firewood, the kilns burn at lower temperatures to melt the lead away. (Roberts, 2019)

Participant testimonies may help to explain why people are not very willing to be convinced by individualizing campaigns. The participants themselves have much more complex views regarding what outsiders see as health hazards. Soda and lead-glazed plates may make you sick, but that is often not immediate or certain. What is certain is that both are a part of the neighbourhood inhabitants' ways of showing affection for each other. Soda and sweets make you fat, but 'thinness is not necessarily to be striven for where food is love and fat is a sure sign of existence' (Roberts, 2015b). Even the toxic smell of the sewage has its benefits; it protects Colonia Periférico against police violence (Roberts, 2017). Disability theory and non-ideal theory may help to understand the complex view of inhabitants regarding environmental hazards as both harmful and protective.

Conclusion

Striving towards epigenetic justice in Colonia Periférico does not involve quick fixes. In fact, ascribing responsibility and finding ways to hold agents accountable will never be a straightforward effort in the face of the complex web of epigenetic mechanisms and environmental factors. In this chapter, I contribute to the debates on the potential role of forward-looking collective responsibility in the context of epigenetics. I show how the concept can be useful when connected with a clear aim and backed up by appropriate normative commitments. In such situations, ascriptions of FLCR may be justified on both forward-looking grounds (pertaining to consequences) and more backward-looking claims.

Finally, distributing FLCR also means looking at one's own role or place in structures and collectives that are either related to existing health disparities in some way or may help to improve them. We do not need to be public health experts or CEOs of polluting corporations to do so. As I have pointed out, epigenetic justice is intricately connected with well-known disparities and inequalities. Working towards women's rights, eradicating poverty and increasing disability justice are important goals in themselves, but may also bring us closer to a more equitable epigenetic future. This may strike some as overly demanding, but as Young puts it 'in a world with significant and multiple structural injustices, people's responsibility in relation to those injustices can and should appear to be too much to deal with' (Young, 2011, p 123). Indeed, epigenetic injustices are so pervasive and structural that no individual or collective can address all of them. This should not prevent us from taking action.

Notes

[1]   However, as Martin Sand helpfully pointed out to me, not all accounts of FLCR include direct implications for what desirable state should be brought about (see, for example, Held, 2006).

[2]   For an overview of debates on collective responsibility, see Smiley (2022). For an explanation of some accounts defending the cogency of the concept, see Sand (2018).

[3]   Eliminativism is the philosophical view that 'we should eliminate our belief in responsibility and our corresponding responsibility practices (blame, praise)' (Ciurria, 2019, p 233). A prominent defender of eliminativism is Bruce Waller (Waller, 2011).

[4]   This notion is inspired by Sally Haslanger (see, for example, Haslanger, 2006).

[5]   Complex causality relationships do not mean that epigenetic knowledge is not actionable at all. Even if the precise extent to which a certain factor contributes to an outcome may be very hard to determine, it may nonetheless be clear that the factor or agent contributed to some extent. Moreover, epigenetic epidemiology can discover tendencies in populations and detect significant statistical associations (Santaló and Berdasco, 2022).

[6]   As various authors have noted, contrasting evidence of environmental exposure with lived experience may raise tensions between identifying harmful (environmental) influences on the one hand and not wanting to attach negative value to the bodies shaped by such influences on the other (Kafer, 2013; Clare, 2017; Bretz, 2020; Saulnier, 2020).

7   This general statement does not deny that defining the degree of complicity of an agent is sometimes (nearly) impossible (Posner and Sunstein, 2007, p 1597).

8   This may be compared with Hart's 'role responsibility' (Hart, 2008) or Miller's emphasis on community membership as a potential way to identify remedial responsibility to come to the aid of those who may need help (Miller, 2007).

References

Allen, A. (2021) 'Feminist perspectives on power', in E.N. Zalta and U. Nodelman (eds), *The Stanford Encyclopedia of Philosophy* [online], 28 October. Available from: https://plato.stanford.edu/entries/feminist-power/ [Accessed 27 March 2022].

Barnes, E. (2014) 'Valuing disability, causing disability'. *Ethics*, 125(1): 88–113. https://doi.org/10.1086/677021.

Björnsson, G. and Brülde, B. (2017) 'Normative responsibilities: structure and sources', in K. Hens, D. Horstkötter and D. Cutas (eds), *Parental Responsibility in the Context of Neuroscience and Genetics*, Berlin: Springer, pp 13–33.

Bretz, TH. (2020) 'Discussing harm without harming: disability and environmental justice', *Environmental Ethics*, 42(2): 169–87.

Chiapperino, L. (2018) 'Epigenetics: ethics, politics, biosociality', *British Medical Bulletin*, 128(1): 49–60.

Chiapperino, L. (2020) 'Luck and the responsibilities to protect one's epigenome', *Journal of Responsible Innovation*, 7(suppl 2): S86–106.

Ciurria, M. (2019) *An Intersectional Feminist Theory of Moral Responsibility*. New York: Routledge.

Clare, E. (2017) *Brilliant Imperfection: Grappling with Cure*, Durham, NC: Duke University Press.

Dupras, C. and Ravitsky, V. (2016) 'The ambiguous nature of epigenetic responsibility', *Journal of Medical Ethics*, 42(8): 534–41.

Dupras, C., Saulnier, K.M. and Joly, Y. (2019) 'Epigenetics, ethics, law and society: a multidisciplinary review of descriptive, instrumental, dialectical and reflexive analyses', *Social Studies of Science*, 49(5): 785–810.

Grossi, É. (2020) 'New avenues in epigenetic research about race: online activism around reparations for slavery in the United States', *Social Science Information*, 59(1): 93–116. https://doi.org/10.1177/0539018419899336.

Hart, H.L.A. (2008) *Punishment and Responsibility: Essays in the Philosophy of Law*, Oxford: Oxford University Press.

Haslanger, S. (2006) 'What good are our intuitions: philosophical analysis and social kinds', in *Proceedings of the Aristotelian Society Supplementary*, Vol. 80, pp 89–118.

Hedlund, M. (2012) 'Epigenetic responsibility', *Medicine Studies*, 3(3): 171–83. https://doi.org/10.1007/s12376-011-0072-6.

Heijmans, B.T. and Mill, J. (2012) 'Commentary: the seven plagues of epigenetic epidemiology', *International Journal of Epidemiology*, 41(1): 74–8.

Held, V. (2006) *The Ethics of Care: Personal, Political, and Global*, Oxford: Oxford University Press.

Hens, K. and Van Goidsenhoven, L. (2017) 'Autism, genetics and epigenetics: why the lived experience matters in research', *BioNews*, 929: 1–3.

Huang, J.Y. and King, N.B. (2018) 'Epigenetics changes nothing: what a new scientific field does and does not mean for ethics and social justice', *Public Health Ethics*, 11(1): 69–81.

Kafer, A. (2013) *Feminist, Queer, Crip*, Bloomington, IN: Indiana University Press.

Katz, MB. (2018) 'The biological inferiority of the undeserving poor', in O.K. Obasogie and M. Darnovsky (eds), *Beyond Bioethics: Toward a New Biopolitics*, Berkeley, CA: University of California Press, pp 17–31.

Mansfield, B. and Guthman, J. (2015) 'Epigenetic life: biological plasticity, abnormality, and new configurations of race and reproduction', *Cultural Geographies*, 22(1): 3–20.

McGuinness, D., McGlynn, L.M., Johnson, P.C.D., MacIntyre, A., Batty, G.D., Burns, H., et al (2012) 'Socio-economic status is associated with epigenetic differences in the PSoBid cohort', *International Journal of Epidemiology*, 41(1): 151–60.

Meloni, M. and Müller, R. (2018) 'Transgenerational epigenetic inheritance and social responsibility: perspectives from the social sciences', *Environmental Epigenetics*, 4(2): dvy019.

Mill, J. and Heijmans, B.T. (2013) 'From promises to practical strategies in epigenetic epidemiology', *Nature Reviews Genetics*, 14(8): 585–94.

Miller, D. (2007) *National Responsibility and Global Justice*, Oxford: Oxford University Press.

Mills, C.W. (2005) '"Ideal theory" as ideology', *Hypatia*, 20(3): 165–83.

Neuhäuser, C. (2014) 'Structural injustice and the distribution of forward-looking responsibility', *Midwest Studies in Philosophy*, 38: 232–51.

Nuti, A. (2019) *Injustice and the Reproduction of History: Structural Inequalities, Gender and Redress*, Cambridge: Cambridge University Press.

Posner, E.A. and Sunstein, C.R. (2007) 'Climate change justice', *Georgetown Law Journal*, 96: 1565.

Radzik, L. (2014) 'Historical memory as forward-and backward-looking collective responsibility', *Midwest Studies in Philosophy*, 38: 26–39.

Roberts, E.F.S. (2015a) 'Bio-ethnography: a collaborative, methodological experiment in Mexico City', *Somatosphere* [blog], 26 February. Available from: http://somatosphere.net/2015/bio-ethnography.html/ [Accessed 26 March 2022].

Roberts, E.F.S. (2015b) 'Food is love: and so, what then?', *BioSocieties*, 10(2): 247–52.

Roberts, E.F.S. (2017) 'What gets inside: violent entanglements and toxic boundaries in Mexico City', *Cultural Anthropology*, 32(4): 592–619.

Roberts, E.F.S. (2019) 'Bioethnography and the birth cohort: a method for making new kinds of anthropological knowledge about transmission (which is what anthropology has been about all along)', *Somatosphere* [blog], 19 November. Available from: http://somatosphere.net/2019/bio ethnography-anthropological-knowledge-transmission.html/ [Accessed 26 March 2022].

Rosen, E.D., Kaestner, K.H., Natarajan, R., Patti, M.-E., Sallari, R., Sander, M. and Susztak, K. (2018) 'Epigenetics and epigenomics: implications for diabetes and obesity', *Diabetes*, 67(10): 1923–31.

Rothstein, M.A. (2013) 'Legal and ethical implications of epigenetics', in R.L. Jirtle and F.L. Tyson (eds), *Environmental Epigenomics in Health and Disease*, Berlin: Springer, pp 297–308.

Sand, M. (2018) 'Collective and corporate responsibility', in *Futures, Visions, and Responsibility: An Ethics of Innovation*, Berlin: Springer, pp 201–39.

Santaló, J. and Berdasco, M. (2022) 'Ethical implications of epigenetics in the era of personalized medicine', *Clinical Epigenetics*, 14(1): 44.

Saulnier, K. (2020) *Epigenetics and the Disabled Self: Finding an Ethical Narrative in the Shifting Boundaries of Molecular Science*. Available from: https://escho larship.mcgill.ca/downloads/mw22vb076 [Accessed 10 March 2022].

Saulnier, K.M. and Dupras, C. (2017) 'Race in the postgenomic era: social epigenetics calling for interdisciplinary ethical safeguards', *The American Journal of Bioethics*, 17(9): 58–60.

Saulnier, K., Berner, A., Liosi, S., Earp, B., Berrios, C., Dyke, S., et al (2022) 'Studying vulnerable populations through an epigenetics lens: proceed with caution', *Canadian Journal of Bioethics*, 5(1): 68–78.

Senut, M.-C., Cingolani, P., Sen, A., Kruger, A., Shaik, A., Hirsch, H., et al (2012) 'Epigenetics of early-life lead exposure and effects on brain development', *Epigenomics*, 4(6): 665–74.

Smiley, M. (2014) 'Future-looking collective responsibility: a preliminary analysis', *Midwest Studies in Philosophy*, 38(1): 1–11. https://doi.org/ 10.1111/misp.12012.

Smiley, M. (2022) 'Collective responsibility', in E.N. Zalta and U. Nodelman (eds), *The Stanford Encyclopedia of Philosophy* [online], 19 December. Available from: https://plato.stanford.edu/entries/collective-responsibil ity/ [Accessed 12 February 2023].

Sullivan, S. (2013) 'Inheriting racist disparities in health: epigenetics and the transgenerational effects of white racism', *Critical Philosophy of Race*, 1(2): 190–218.

Téllez-Rojo, M.M., Hernández-Avila, M., González-Cossío, T., Romieu, I., Aro, A., Palazuelos, E., et al (2002) 'Impact of breastfeeding on the mobilization of lead from bone', *American Journal of Epidemiology*, 155(5): 420–8.

Vears, D.F. and D'Abramo, F. (2018) 'Health, wealth and behavioural change: an exploration of role responsibilities in the wake of epigenetics', *Journal of Community Genetics*, 9(2): 153–67.

Waller, B.N. (2011) *Against Moral Responsibility*, Cambridge, MA: MIT Press.

Wang, T., Zhang, J. and Xu, Y. (2020) 'Epigenetic basis of lead-induced neurological disorders', *International Journal of Environmental Research and Public Health*, 17(13): 4878. https://doi.org/10.3390/ijerph17134878.

Warin, M., Kowal, E. and Meloni, M. (2020) 'Indigenous knowledge in a postgenomic landscape: the politics of epigenetic hope and reparation in Australia', *Science, Technology & Human Values*, 45(1): 87–111.

Young, I.M. (2011) *Responsibility for Justice*, Oxford: Oxford University Press.

3

Luck, Epigenetics and the Worth of Collective Agents

Luca Chiapperino and Martin Sand

Introduction

The possibility of describing the effects of lifestyles and/or environmental exposures through measures of epigenetic modifications has prompted a prolific debate around the responsibility claims attached to this knowledge. Social sciences and humanities scholars have formulated several critiques of individual claims regarding uses of epigenetic information for responsibility attribution (Hedlund, 2012; Dupras and Ravitsky, 2016; Chiapperino, 2018; Meloni and Müller, 2018; Bolt et al, 2020). Specifically, critiques have focused on the limitations of two intertwining responsibility claims (Vincent, 2011): one in terms of accountability for damaging one's own epigenome (liability or backward-looking responsibility), and another one highlighting prospective duties to protect it (remedial or forward-looking responsibility). Aside from these critiques, moral luck has been introduced as another challenge of such responsibility claims (Chiapperino, 2020). The long-standing debate on luck in moral philosophy (Williams, 1982; Nagel, 1991; Statman, 1993) has examined the effect that factors beyond one's control have on the justification and cogency of normative claims such as responsibilities. The challenge of luck for moral intuitions concerning responsibility resonates well with a consideration of the epigenome's complexity and stochasticity (Panzeri and Pospisilik, 2018). Unlike other critiques of responsibilities grounded on epigenetics (see Hedlund, 2012), considerations of luck question the causality conditions of these responsibility claims. Not only is it difficult to disentangle whether an epigenetic modification is solely due to lifestyle, environmental stimuli, genetic differences or stochasticity, but the complex causation of epigenetic modifications also calls into question an agent's capacity to affect this course of action. Considering these factual considerations, previous work has challenged the idea that individuals really affect their epigenome and that they can therefore be held responsible for past behaviours and/or future actions remedying these health risk factors (Chiapperino, 2020).

However, such a criticism based on luck also dramatically jeopardizes the possibility to meaningfully ascribe responsibilities to prevent or correct epigenetic harms to collective agencies (for example, the state, corporations, public health agencies). Collectives are also subject to circumstances, conditions and vagaries in the outcomes of actions, raising the problem of moral luck (Chiapperino, 2020). But does considering luck in the normative uptake of epigenetics leave us without any notion of epigenetic responsibility altogether? This chapter aims to explore whether any residual collective epigenetic responsibility remains after taking into account the challenge of moral luck. Both ordinary language and the social function of collective responsibilities call for an effective societal uptake of epigenetic knowledge. However, this requires an appropriate language of responsibility. Our goal here is to specify in what salient ways collective agencies should be blamed for failing to prevent, remedy or be accountable for epigenetic predispositions to health problems caused by socio-environmental exposures. To this purpose, we develop a different approach to mitigate the effects of moral luck on (at least) a residual teleological/role version of responsibility. The model draws on notions of aretaic blame (Cheng-Guajardo, 2019) to argue that collective (for example, corporate, state or public health) commitments (or failures to commit) to the protection of our health are crucial for moral evaluation of the worth of these collective agents. This shall be taken to imply a preoccupation with the interaction between health and the environment insofar as this is mediated by the epigenome. As distinguished from a strong version of moral responsibility, this approach embraces a moral life of epigenetic knowledge that considers the complex circumstances, social processes, indirect agencies, intricate causalities and transformative opportunities characterizing the roles of both collective agents and the epigenome in shaping health trajectories. We first provide an overview of how evidence of epigenetic modifications is tied in the literature to questions of individual and collective responsibility. We then discuss how luck challenges the attribution of such responsibilities. We conclude by offering a resolution to this challenge, focusing on an assessment of collective agents' moral worth as residual collective responsibility.

Epigenetics and responsibility claims: strands of criticism

Commonly studied epigenetic modifications, such as DNA methylation, are currently emerging as accessible biomarkers of the effects of lifestyle and/or environmental exposures on health (Guerrero-Preston et al, 2011). Global and gene-specific methylation patterns have been associated with different individual behaviours, social conditions, environmental exposures and lifestyles. Although the causal implication of epigenetic modifications in disease aetiology is still debated (Shanthikumar et al, 2020), several researchers have underlined the practical utility of this information (Cooney, 2007; Fiorito et al, 2019). Epigenetic modifications are not only regarded as a footprint of experiences, environmental exposures and life trajectories, but allegedly also offer an insight into the mechanisms of health and disease (Cavalli and Heard, 2019). In a nutshell, researchers invest this information with the potential to both illuminate the mode of action of exposures (chemical, social, lifestyle, and so on) on the body (how the body responds to environmental cues) (Jeremias et al, 2020), and offer actionable mechanisms of disease 'that can lead to better prediction, prevention, treatment, and policy' (Ladd-Acosta and Fallin, 2019, p 2).

This dimension of actionability of epigenetic information has been the subject of substantial scrutiny. While the potential of policies focusing on social and environmental interventions based on epigenetics has been acknowledged (Chiapperino and Testa, 2016; Chung et al, 2016), it remains unclear how to incorporate epigenetic information into normative discourses of responsibility. What if epigenetics becomes politicized as the science of desert and accountability in healthcare, as well as responsibility for protecting the epigenome and health (Hedlund, 2012; Loi et al, 2013; Rothstein, 2013; Chiapperino and Testa, 2016; Bolt et al, 2020)? This debate has been particularly prolific because epigenetic knowledge touches upon standard conditions for models of both backward- and forward-looking responsibility (Pettit, 2007; Aristotle, 2009; Vincent, 2011; Talbert, 2019). First, epigenetics allegedly brings to light the causal connections between a particular agent, or a given set of actions (for example, lifestyles, environmental exposures), and a certain responsibility-relevant outcome with regard to responsibility (for example, one's health condition). For any claim of (backward- and forward-looking) responsibility, it is usually a necessary condition that the agent has causally contributed to an outcome or can contribute to remedying it. Epigenetic marks of past behaviours epitomize these causal intuitions around responsibility, even though they are far from doing so without any doubt (see below).

Second, another component of moral conceptions of responsibility is the so-called voluntary condition, which postulates that the agent may be judged responsible if the action under scrutiny was voluntary, that is the agent had control over whether the action/outcome emerged as this was neither a necessity nor a random event (see Talbert, 2019 for an introductory overview of various approaches to the voluntariness condition).

Finally, epigenetic information relates to the moral intuition connecting responsibility with the degree of knowledge that we hold about our actions and their consequences: the more we know about what is at stake, the more we can be held responsible for our actions, or for remedying a state of affairs. Known as the epistemic condition (Pettit, 2007; Aristotle, 2009), this point is particularly relevant to the ethical scrutiny of epigenetic knowledge. Does this novel information about the impact of one's actions and/or life conditions over health 'make a change in degree' (Hedlund, 2012, p 178) in the responsibilities that individuals hold to protect their health? Does this open new questions of responsibility in light of previously unknown multigenerational effects of unhealthy behaviours (Chadwick and O'Connor, 2013)?

Critical studies of epigenetics provide a rich normative basis for deconstructing claims relating to both backward- and forward-looking individual responsibilities for protecting one's epigenome (reviewed in Chiapperino, 2018; Dupras et al, 2019; Santaló and Berdasco, 2022). Primarily, and following an extensive body of scholarship on responsibility in relation to health (Minkler, 1999; Resnik, 2007; Buyx, 2008; Brown, 2013; Voigt, 2013), scholars have questioned the voluntariness and cognizance conditions of backward-looking claims towards epigenetically grounded accountability for unhealthy lifestyles and behaviours (for example, Bolt et al, 2020). In a seminal article, political scientist Maria Hedlund pointed to the 'circumstances that to varying extent constrain individual choice' (Hedlund, 2012, p 179) to undermine claims of intentionality, voluntariness and capacity around responsibility concerning our epigenome (see also Chapter 6). In her view, these conditions rarely apply, as the involved parties are constrained by unequal social and economic structures. Even if one conceded that the epigenome highlights previously unknown mechanisms linking lifestyles, environmental exposures and our bodies, it would be excessive to claim that lifestyle behaviours result from individual deliberate and knowledgeable choices regarding a course of action. Individuals seldom have (in a morally relevant sense) control over their lifestyle behaviours as well as the (epigenetically mediated) outcomes they bring about. Instead, those behaviours stem from an intricate web of social structures and influences that 'strike unevenly' (Hedlund, 2012, p 179) in our societies, and thus unevenly hamper individual capacities to take full responsibility for their consequences, or for correcting them.

In the face of these criticisms of the voluntariness and epistemic conditions for epigenetic responsibilities (both backward- and forward-looking), several scholars have suggested that the responsibilities for protecting the population's epigenome should largely be ascribed to collective agents. As famously argued by Hedlund, epigenetic knowledge 'calls attention to the role of structural conditions, which as well could give rise to a focus on *the role of society and the state to protect and care for health and wellbeing of individuals, present and in the future*' (Hedlund, 2012, p 181; emphasis added). As many of the contributions to this volume testify, the fact that large structural social configurations influence health, patterns of environmental exposures or individual behaviours – all processes with distinct epigenetic effects on health – demands collective, rather than individual, action to account for and/or remedy this state of affairs. Critical and cautious voices notwithstanding (Dupras and Ravitsky, 2016; Hens, 2017; Huang and King, 2018), an overarching consensus exists as to a normative translation of epigenetics promoting 'a forward-looking approach that calls for collective responsibility' (Pentecost and Meloni, 2018, p 62).

Does luck undermine collective epigenetic responsibilities?

Other critics have taken issue with the actionability of epigenetic information (or lack thereof). Another critique of these claims is, in other words, asking whether they meet the causal and epistemic conditions of responsibility. Does epigenetic knowledge offer novel avenues for taking control of one's health? And even if lifestyles and/or exposures are implicated in disease through epigenetic mechanisms, does this information really heighten our knowledge and inform action? Previous work (Chiapperino, 2018; Chiapperino, 2020) has deconstructed claims of the backward-looking type of responsibility by pointing to the nature of the epigenome and epigenetic mechanisms, as well as to the ways causal claims are discussed in the biomedical debate internal to environmental epigenetic and epigenetic epidemiology (Heijmans and Mill, 2012; Mill and Heijmans, 2013; Mitchell, 2018). In risk assessment contexts, it is still a 'fundamental challenge' to identify 'measurable causal relationships between epigenetic modifications and health outcomes' (Angrish et al, 2018). The existing scientific evidence reporting the epigenetic effects of past individual exposures, habits, life conditions and psychosocial factors on these mechanisms lacks a clear understanding of the causal connections required to establish responsibility. The relationship between epigenetic modifications, gene expression and resulting health phenotypes is complex. There is still limited knowledge of how the epigenome functions in different genomic contexts (for example, tissue types) (Jones, 2012; Birney et al, 2016). But also, an organism's complex traits (such as most diseases) are hard to predict from epigenetic parameters alone. Phenotypes result from multicausal relationships that flow in multiple directions among genetic, epigenetic, cellular, organismic and environmental factors. These processes are also heavily affected by developmental trajectories, and are partly the result of stochasticity in determining genomic regulatory outcomes and phenotypic effects (Panzeri and Pospisilik, 2018). Nowadays, epigenetic stochastic variance is recognized as an important contributor to phenotypic variation within a population (Peaston and Whitelaw, 2006; Allis and Jenuwein, 2016). Stochastic changes in DNA methylation that may be transmitted from one generation to the next, also called 'spontaneous epimutations', have been studied for years in plant species (reviewed in Johannes and Schmitz, 2019) but remain a puzzle for scientists studying the impact of epigenetics on human disease and inheritance (Biwer et al, 2020). Finally, epigenetic evidence does not fully support the idea of reversibility, especially in cases where developmental dynamics have contributed to the establishment of a disease phenotype. Plasticity in adulthood is only residual, resulting in limited possibilities for individuals to revert aberrant metabolic processes and reduce disease progression through actions whose effects are mediated by the epigenome (Panzeri and Pospisilik, 2018).

These caveats are necessary to accurately interpret how this biological information affects the impact of agents on the body, ageing and disease (through the epigenome). There is little possibility of adjudicating whether an epigenetic modification is due solely to lifestyles, environmental stimuli, genetic differences or stochasticity. Similarly, it is also challenging to disentangle to what extent an outcome is due to any of these factors. A different and related version of this critique can be formulated concerning the duty to adjust one's behaviours or take a course of action to repair or remedy to aberrant epigenetic predispositions towards disease (forward-looking responsibilities). Dupras and Ravitsky (2016) have taken issue with these claims based on similar epistemic considerations about the epigenome. It would also be difficult to enact such responsibility claims prospectively in an informed way as the complexity of the epigenome undermines any definition of 'epigenetic normality' (Dupras and Ravitsky, 2016, p 536): this is highly contextual, being relative to a unique assessment of an organism's genetic, epigenetic, environmental and developmental trajectory, as well as open to luck and stochasticity. Thus, what 'healthy' behaviours and what specific epigenetic effects should one strive for? Furthermore, one may also add that the stochasticity of epigenetic effects questions an agent's capacity to causally affect this course of action: can individuals really affect their epigenome when the outcome of their actions lies beyond their control?

A previous contribution to this debate (Chiapperino, 2020) formulated a critique of both retrospective and prospective epigenetic responsibilities under the banner of the renowned philosophical problem of moral luck (Williams, 1982; Nagel, 1991; Statman, 1993). Standard notions of responsibility are at odds with the idea that we might be held responsible for the epigenetic effects of our behaviours, lifestyles or exposures, if it cannot be proved that we have willingly and intentionally brought them about. The considerations previously discussed concerning how stochasticity and luck affect our epigenome as a result of behavioural and environmental factors raise precisely this challenge to standard notions of responsibility; a paradox that may be well apprehended in terms of moral luck. On the one hand, it may be argued that the epigenome's complexity and stochasticity indicate that we cannot be held morally responsible for epigenetic modifications because we do not actually cause and control them, requiring the admission that the pervasiveness of luck in our lives (and epigenomes) dramatically undermines responsibility for these effects. On the other hand, those considerations might be taken as a reason for shielding our judgements of someone's responsibility for these effects from luck. But then we would end up dramatically restricting the ground for attributability and ownership of these actions. Luck (as stochasticity, but not exclusively: see Chiapperino, 2020) is in fact ubiquitous in the way complex metabolic phenotypes emerge, to the extent that excluding luck would leave little scope for responsibility to apply.

Let us spell out how luck provides another source for critique against responsibility claims around the epigenome. Luck wears out the moral concept of responsibility in relation to biological factors regarding the effects resulting from one's action. To paraphrase philosopher Thomas Nagel, there is luck in 'the way things turn out' in the epigenome and its role in health (Statman, 1993, p 61). Our epigenome is characterized by environmental plasticity, individual variability and a general indeterminacy of change–effect mechanisms. By putting resultant luck into the picture, we are left with a very different understanding of the moral cogency of claims towards (epigenetic) responsibility. Specifically, one may highlight three potential sources of resultant luck for a given agent and an epigenetic outcome.[1] First, the outcome itself of lifestyles and/or exposures may occur or fail to occur. It is far from clear whether specific lifestyles or environmental exposures produce aberrant epigenetic predispositions (other factors, including stochasticity, may bring them about). Second, it is unclear whether the agent may bring about a specific outcome at the level of the epigenome or fail to bring it about; this is conditional on factors that are not affected by the actions themselves (for example, temporalities, genetic variability, stochasticity). Third, and most relevant to forward-looking claims, it is uncertain whether there is a 'right' way in which an agent can bring about the outcome. Epigenomes change during the course of development, as a result of individual genetic differences and due to stochasticity, which defies precise determination of what behaviours are conducive to health and should be pursued. As argued elsewhere, the 'success – and, perhaps, also the praise or blame – attached to these exercises of responsibility seems to be the result of much more than behaviours, choices and actions of the concerned agents' (Chiapperino, 2020, p 8).

However, the problem with this luck-based critique is that it also has a dramatic impact on the assumption that there are collective epigenetic responsibilities. A luck-based approach highlights the vulnerabilities, circumstances and uncertainties that call into question the coherence of agents to whom responsibility is ascribed. Similar concerns may be relevant for the ascription of responsibility to collective agents (Lewis, 1948; Feinberg, 1968; French, 1984; Arendt, 1987; Smiley, 2022; see also Chapter 2). A previous paper pointed out how luck suggests that collective epigenetic responsibilities 'fail to be an obvious alternative normative construct to their individualistic counterparts' (Chiapperino, 2020, p 2). Even if we hold a coherent view of collective agents as the bearers of responsibility, these are in fact no less exposed to luck than individual agents, in ways that would temper attributions of responsibility. In fact, it may be questioned whether their actions to prevent, neutralize or reverse potentially damaging epigenetic effects are reasonably the target of responsibility judgements. A series of intertwining factors outside policy control or corporate agency may arguably be invoked to deflect these claims. Whether an individual or a group is predisposed or vulnerable to the epigenetic drivers of complex diseases results from many factors, including the stochastic or highly contextual dependency of epigenetic mechanisms (Panzeri and Pospisilik, 2018). Whether an individual is likely to be exposed more than another, and what the harmful consequences for that specific individual may be, are all outcomes that are not strictly under the control of these collective agents. These consequences may depend on historically distant actions, practices and inequalities that persist, or even unique combinations of biological and/ or environmental determinants of health for the individual in question. As argued by environmental justice scholar Levente Szentkirályi (in a separate yet contiguous context), it is partly a matter of luck 'whether or not emitters who create uncertain threats are culpable' of anything, as much as it is a matter of luck 'whether some may be injured by their actions' (Szentkirályi, 2020, p 8). Given the inability to ascertain whether environmental exposures, social structures or life contexts do cause aberrant epigenetic predispositions to disease, the responsibility of collective agents under such circumstances appears to diminish. But does this mean that collective agents are blameless under all circumstances for not taking (backward- and forward-looking) responsibility for the proliferation of epigenetic predispositions towards disease among their populations?

Moral worth and the residual responsibilities of collective agents

One problem with the criticism from luck and its significance for ascribing collective responsibility is that it undermines the possibility of ascribing blame to collective agents for failing to remedy or prevent health risks and epigenetic predispositions to diseases. While the coherence and moral cogency of collective agents may be the focus of a prolific philosophical debate, the public expression of blame also plays an important social function. The state and its public health branches are expected to contribute to the overall welfare of citizens (Pettit, 2007). Whether philosophically cogent or not, in reality, corporations are the target of moral blame when they fail to respond to the needs of society or fail to benefit society through their actions. Such a 'collectivist' position on responsibility is attractive and feeds into common moral intuitions about the state, public health agencies or corporations: they should be held accountable for their actions, especially when they perform or fail to perform some of them. Otherwise stated, there can be no denying that 'we lose something important' (Cheng-Guajardo, 2019, p 295) if we fail to account for typical moral sentiments (for example, disappointment, expectation, blame) that are commonly oriented towards collective agents.

Following the suggestion of moral theorist Luis Cheng-Guajardo[2], our intention in this section is to formulate an approach for blaming collective agents and holding them residually responsible for protecting our epigenomes in the face of the challenge that luck poses for the coherence of moral theories of collective responsibility. By using the term 'residually', we intend to underline a distinction between full responsibility claims (for example, those that meet the criteria of standard moral conceptions of responsibility) and weak or expansive uses of the term (see Wolf, 2001). These can encompass: (1) the pragmatic foreshortening of responsibility due to lack of insight into whether responsibility conditions are met (such a foreshortening therefore lacks an assessment of the coherent agency of collective agents, see Sand, 2018, chapter 6), (2) attributions of responsibility based on the role that collective agents can play more than their actual ownership of the actions for which responsibility is sought (Pettit, 2007), and (3) exercises of responsibility that exceed the challenges of luck and objective responsibilities for reasons of virtue, solidarity and the moral community (Wolf, 2001). The model we propose relies on the idea that the moral 'worth' of collective agents can justify responsibility claims of the third type to prevent the health effects of structural social conditions or exposures, including adverse epigenetic modifications.

Our approach draws from notions of aretaic blame (Watson, 1996; Cheng-Guajardo, 2019) that emphasize blameworthiness as a teleological failure, or the failure to meet one's purpose, objective and goals. Failing/succeeding in the realization of one's *telos* in fact reveals something about oneself, namely that one achieves what one is well-positioned to achieve, that one cares about others, that one participates for the benefit of a community of shared values and goals beyond mere obligations and bounded responsibilities. This suggests that collective commitments (or failures to commit) to the protection of our health and epigenomes have a deep ethical import for the evaluation of these collective agents vis à vis their *telos*. Given the intricacies of attribution, aretaic blame does not consider individual actions of collective agents. Rather it takes those agents as temporally extended entities, whose various ways of affecting society lead to the emergence of patterns that allow identification of their dispositions and traits. Without the need for a full notion of moral responsibility, an aretaic appraisal of collective agents involves a weaker attribution of responsibility to act on the social structures and environmental factors that contribute to the epigenetic dysregulation of bodies and the occurrence of disease in a given population.

Let us begin with the problem that arises from the difficulty of disentangling the relationship between certain undertakings (damaging or protecting the epigenome of a population) and a given (successful or not) outcome from the perspective of luck. As we have shown, these outcomes depend on stochasticity, multiple causations and the complexity of epigenetically driven phenotypic variation. Generally speaking, resultant luck refers to the outcome of an agent's acts, characterizing these results as being 'beyond the agent's control, or not fully within the agent's control' (Sartorio, 2012, p 63). Based on this view, it seems unreasonable to hold agents morally responsible for some of those results. For any course of action aiming to prevent an epigenetic effect, one could plausibly find in fact an alternative course of action that differs only for factors that may be bona fide taken as luck (for example, genetic contribution to an epigenetic effect and/or to the resulting phenotype, stochasticity, an environmental confounder). Hence, can collective agents be praised for bringing about a beneficial outcome, or blamed for failing to produce courses of action beyond their control?

As argued by philosopher John Greco in a seminal article on luck and responsibility (Greco, 1995), this formulation of the paradox of luck may be solved in two main ways. The first rescues causation and control from the challenge that resultant luck seems to raise. This famous solution to the paradox of luck portrays the problem as being only an apparent one (Zimmerman, 1987). There is more than one sense in which an agent can willingly cause an action. Hardly anyone would think that the state, or a public health agency, is responsible, for instance, for the outcome of policies preventing aberrant epigenetic modifications due to environmental exposures. What one expects from any agent is to exercise the 'restricted' control (Zimmerman, 1987, p 376) that they can exert to remove the sources of these exposures, and not to control all those events on which their epigenetically mediated effects over health depend. As argued elsewhere, this solution, although appealing in several respects, may nonetheless be only partial (Sand and Klenk, 2021). Zimmerman's critique of the luck paradox restores standard intuitions on the control condition for responsibility: agents whose undertakings are susceptible to luck can nonetheless be morally responsible for wanting to bring them about (Zimmerman, 2002, p 559; Hanna, 2014). Yet, this critique may only partly solve the problem raised by the normative exercises demanded by epigenetic knowledge. At least to the extent that many of these epigenetic effects are open to multiple causalities and indeterminacy, the challenge from luck is not just about the responsibility-undermining lack of control over the outcomes of such potential policies. The problem also lies in acquiring genuine knowledge of the causal chain of events that brings about a beneficial outcome and choosing a course of action that brings about that outcome. Let us consider an example.

Tests that are used to assess the risk of a family of chemicals such as endocrine disruptors rarely address persistent effects arising from early-life exposures, microdoses or mixtures of these chemicals to which we are exposed on a daily basis (Alavian-Ghavanini and Rüegg, 2018). Most importantly, data on the adverse phenotypic outcomes of exposure to these substances are often absent or there is a lack of evidence of any causal relationship between the adverse outcomes and the exposure to the chemicals in question. There is a growing recognition that, while this is partly a problem of uncertainty (understood as knowledge to be yet produced), it is also difficult to draw definitive conclusions about harm from environmental exposure (for a critique of uncertainties and inaction in environmental risk assessment, see Szentkirályi, 2020). The incorporation of epigenetic endpoints into chemical risk assessment may offer novel mechanistic insights into the modes of action of a substance. However, it does not necessarily provide a more effective characterization of its hazardous properties (Garcia-Reyero and Murphy, 2018). In fact, within regulatory circles, a paradigm shift is often called for, from a hazard-driven risk assessment to one that is exposure-driven (European Commission Directorate-General for Health and Consumers, 2013). This approach focuses on the vulnerabilities that various kinds of factors bring, and suggests switching the focus of responsible agency from assessing harm and risks to a precautionary approach. Epigenetic information here may ultimately increase awareness of the conditions of uncertainty and indeterminacy under which these harmful substances may affect citizens, rather than revealing the deleterious consequences of these exposures that the state, corporations or public health actors are compelled to address. Epigenetic alterations are neither necessary nor sufficient conditions of the possibility of disease, but only factors in a probabilistic estimation of their occurrence. As indeterminate threats and genuinely unforeseeable contributors to an outcome, these exposures and their health consequences offer little foundation for the collective agent's obligation to control them.

This is where the second intuition suggested by Greco (1995) may come in handy. In contrast to Zimmerman, he proposes a solution to the paradox of luck that challenges the assumption that 'moral worth' has to be 'closely tied to one's moral record' (Greco, 1995, p 90). This suggestion, he argues, sets out to counter the way we think about the import of luck on morality. Going back to the example of aberrant exposures mentioned earlier, we consider that luck is responsibility-mitigating (for collective agents) because it makes it difficult for agents to select a course of action and control its outcomes. Physiological traits are the integrated output of many biological and non-biological variables in ways that undermine our ability to predict and bring about an exact outcome from epigenetic information (Panzeri and Pospisilik, 2018). Thus, the vagaries of luck affect our evaluation of a collective agent's actions in ways that make us question the intuitive plausibility of their responsibilities. Greco takes up this problem of recognition, and asks whether the outcomes for which an agent can be rightfully praised or blamed are all that matters for one's moral worth. Giving a negative answer to this question, he suggests that moral worth is also what an agent '*would* voluntarily decide and *would* voluntarily do in a variety of circumstances' (Greco, 1995, p 91, original emphasis) that they may never encounter. This allows him to drive a wedge between the difficulty of making a practical judgement about responsibility and the moral quality of the voluntary and deliberative actions performed by an agent. The worth of corporations creating a hazardous environment, under this account, is independent of their actual record of epigenetically driven effects of exposures on people's health. The failure of a state agency to prevent epigenetically mediated health risks is independent of whether these biological processes actually result in a heightened burden of diseases on the exposed population. Simply put, Greco's point is that the actual results or uncertainties beyond the agent's control may be distinguished from judgements about their moral blameworthiness (see also Jensen, 1984).

But what, then, are the grounds on which to define the moral worth of collective agents such as corporations, the state or public health agencies? What appears to be at stake with regard to moral worth in Greco's account is an agent's fundamental character, fundamental goals and evaluative orientations – the agent's *telos*. Otherwise stated, worth relates here to an appraisal of the function that these agents play in the moral community and society. It involves these collective agents' practical identity, or appeals to their success/failure as contributors to society and its wellbeing. This 'aretaic' perspective has ethical depth in an obvious sense (Watson, 1996; Cheng-Guajardo, 2019). It highlights that blame is attributable to agents when they adopt an end, or commit to a conception of value, that they fail to realize although this was in their power. Without entailing the attribution of full (standard) responsibility, aretaic blame offers a condition to assess residual ways of taking responsibility.

This becomes clearer by using further examples. Consider a corporation that releases endocrine-disruptive chemicals into the environment, which may affect population health through epigenetically mediated mechanisms. Whether they are responsible for harming anyone, or whether they are responsible for remedying that harm, depends on the predispositions of the individuals, or co-occurring vulnerabilities (for example, environmental, biological/genetic). It also depends in part on the stochasticity of the epigenome and complex phenotypic traits. Several of these factors are not within the control or comprehension of the corporation, and therefore luck-based considerations mitigating responsibility apply to a judgement of the liability of the corporation. Aretaic forms of moral evaluation instead re-focus evaluation of the moral worth of the company on whether this hazardous situation, to which they contribute, constitutes the proper functioning of the enterprise. The epigenetic consequences of the omnipresence of the endocrine disruptors in post-industrial societies may therefore not be morally apprehended from mere ascriptions of causal responsibility and the accountability of corporations. Rather, an aretaic assessment of their actions allows us to justifiably blame them in a deep sense for endangering the health of a community. Are these companies failing with regard to their role and contribution to the welfare of society?

Another example could be based on a similar evaluation of public health agencies that fail to remedy these hazards. Again, without the need to hold them morally responsible (in the standard sense) for removing the risks that endocrine disruptors cause to the population (via the epigenome), one may interrogate the function of public health agencies and provide an aretaic evaluation as to whether these agencies should be concerned with this course of action, and, consequently, whether they can be blamed for failing to enact such concern adequately. As examples, one may cite the UK Health Security Agency and the Office for Health Improvements and Disparities – the executive branches of government that have replaced Public Health England in holding the role of promoting the health protection and health improvement of the population in England. Even a cursory look at their websites and statutory documents reveals that these agencies have unique as well as distinctive roles in structuring a collective response to the epigenetically mediated threats to health deriving from exposure to chemicals. The worth of these public health structures is fairly well-defined (one could refer to it as their 'virtues' or *telos*, see Rogers, 2004; MacKay, 2022). The UK Health Security Agency, for instance, aims to offer 'intellectual, scientific and operational leadership' to protect communities from the impact of 'health threats' (UK Health Security Agency, 2023). In this respect, its social function is to spearhead an organized and collective effort in society towards containing and minimizing risks of disease and their distribution. Similarly, the Office for Health Improvements and Disparities focuses on 'improving the nation's health' in ways that level 'up health disparities' and 'break the link between background and prospects for a healthy life' (Allison et al, 2023, p 2). This is a particularly relevant element when thinking about the ethical and political dimensions of environmental epigenetics. Even though this information may not compel public health agencies to remove these exposures, it certainly shows that these exposures (and their ensuing biological vulnerabilities) are unequally distributed in society (Hooten, 2022). Public health agencies therefore have powers and goals that are not shared by other civic organizations or governmental branches of the state and are organized around a referent set of virtues – such as compassion, justice, precaution, solidarity – that define their worth. The 'good of their life', as qualified by moral theorist of virtues Alasdair MacIntyre (2007), heavily depends on pursuing standards of excellence and abiding to certain rules that should encompass accepting a certain, residual responsibility to protect our epigenomes and health. Even if this is not their responsibility in a strong, moral sense of the notion, it may be, at the very least, an important or even central consideration when assessing their worth.

Conclusion

In this chapter, we set out to offer an alternative basis for normative claims concerning epigenetic knowledge as a matter of collective agency. First, we highlighted the fact that the origins of epigenetic variance cannot easily be pinned down to environmental exposures, behaviours, lifestyles or social determinants of health alone. Variation in the distribution of epigenetic marks (and their impact on health) is often unpredictable and is open to the effects of developmental trajectories, stochasticity and genetic susceptibilities (Biwer et al, 2020). Second, we detailed how the implications of such epigenetic variability are also moral: this recognition constitutes a major epistemic and causal limitation to responsibility claims addressed at collective agents (as well as individuals for that matter; Chiapperino, 2020). Stochastic epigenetic variability and multiple causalities in the epigenome preclude blame for socio-environmental conditions that are just risks (among others) of disease, and mitigate the backward-looking responsibility of collective agents (for example corporate, state or public health agencies) for any actual disease outcome emerging from this state of affairs. Roughly put, these agents had no control over the way that things turned out for anyone's health and epigenome. Also, these considerations diminish their blameworthiness for failing to remedy such epigenetic risks, or rather their socio-environmental causes: the conditions of uncertainty, stochasticity and complex causation governing their actions make remedying this state of affairs an outcome beyond their control. Thus, collective (forward-looking) responsibilities also lie beyond their fault.

Contra the idea that luck undermines responsibility *tout court*, we have tried to identify other sources in addition to moral or causal responsibility to justify a residual or weaker epigenetic responsibility for collective agents. In doing so, we have not sought to deconstruct the argument from luck (as many have done already; see endnote 1). Rather, we have developed the idea that the 'problem' of moral luck is a reminder about the impure nature of collective moral agency (Walker, 1991). Collective responsibilities do not fit a perfect match between will, cognizance, choice, action and control over outcomes (Smiley, 2022). These agencies are messy in ways that weaken strict considerations of accountability, desert, or even attributability and duty (Sand, 2018, chapter 6). Epigenetic predispositions and their contribution to resultant luck add another dimension to these complexities of collective responsibilities: these normative claims are often conditional on a contested understanding of collective agents, their coherence, limitations, scope and definition. They are also heavily impacted by the material conditions of the possibility for these agencies to produce their target phenomenon. Simply put, collective agents may have limited control over the health outcomes affected by the epigenome as this is open to several other sources of causality and influence. Our point is therefore to separate moral attributions from the diminishing effects that uncertainties and the indeterminacy of the epigenome have on moral responsibility. It may be more productive to orient our moral evaluations of the state, corporations or public health agencies in relation to the epigenome towards the purpose of these agents in the moral community and society. It is worth asking whether reparation for or protection from the uneven distribution of epigenetic predispositions to disease in society requires attributing full moral responsibilities to these collective agents.

From the perspective of what constitutes the proper functioning of these collective agents, there are reasons to work towards promoting a higher appreciation of the ways that such agents commit to conceptions of value and whether they fail to realize it in the end. Preventive and public health measures addressing health inequalities embodied through the epigenome, or the epigenetic effects of polluted environments, should not require serious threats of harm or even actual harm to demand action. Rather, these may be reframed as duties to exercise due care, which define the collective agent's *telos* and moral worth. It may neither be permissible nor excusable for public health agencies to fail to exercise their powers and goals, which are not shared by other organizations or branches of the state, to maximize pursuit of the referent set of virtues defining their role and worth for society. In other words, the above discussion shows that pragmatism and a defined set of collective goals may be better grounds for action on the heightened health risks from exposures and social structures highlighted by epigenetic evidence. Better, specifically, than the requirement to establish the moral responsibilities of collective agents regarding these effects. In a public health system with limited resources, it may also be necessary to complement the above discussion with a consideration of probability estimates for health outcomes of epigenetic marks, or a justification for the need for more public attention regarding this knowledge, as well as an explanation for the prioritization of interventions targeting epigenetic effects. An argument from luck may only be an entry point on the need for a conceptual sharpening of the role of epigenetics for our moral lives. In turn, our model based on worth may only be a broad sketch of the thick moral reasoning and sophisticated actions required to comprehend these collective risks and the benefits stemming from correcting them. What is certain, however, is that the dominant critique pitting collective epigenetic responsibilities against individual ones may not really deliver on such an objective. This chapter suggests that one possible explanation for this is that simply pinning collective responsibilities against individual ones may be missing its target.

Notes

[1] See Sartorio (2012) for the philosophical specification of these dimensions of resultant luck.

[2] In contrast to the author and other contributors to this debate, we do not engage with the metaphysical question on the distinctive traits that make collective agencies autonomous from the agency of their members or lack thereof (Velasquez, 1983). Chapter 2 in this volume offers a rich analysis of the moral cogency of collective responsibilities.

Contributor statement

L.C. is the leading author in the conception and writing of this chapter. M.S. critically reworked the text for important intellectual content. The chapter's revision and finalization have been thoroughly collegial.

References

Alavian-Ghavanini, A. and Rüegg, J. (2018) 'Understanding epigenetic effects of endocrine disrupting chemicals: from mechanisms to novel test methods', *Basic & Clinical Pharmacology & Toxicology*, 122: 38–45. https://doi.org/10.1111/bcpt.12878.

Allis, C.D. and Jenuwein, T. (2016) 'The molecular hallmarks of epigenetic control', *Nature Reviews Genetics*, 17: 487–500. https://doi.org/10.1038/nrg.2016.59.

Allison, R., Roberts, D.J., Briggs, A., Arora, S., and Anderson, S. (2023) 'The role of health protection teams in reducing health inequities: findings from a qualitative study', *BMC Public Health*, 23: 231. https://doi.org/10.1186/s12889-023-15143-7.

Angrish, M.M., Allard, P., McCullough, S.D., Druwe, I.L., Chadwick, H.L., Hines, E., et al (2018) 'Epigenetic applications in adverse outcome pathways and environmental risk evaluation', *Environmental Health Perspectives*, 126: 045001. https://doi.org/10.1289/EHP2322.

Arendt, H. (1987) 'Collective responsibility', in S.J.J.W. Bernauerz (ed), *Amor Mundi: Explorations in the Faith and Thought of Hannah Arendt*, Dordrecht: Springer, pp 43–50.

Aristotle (2009) *The Nicomachean Ethics*, Oxford: Oxford University Press.

Birney, E., Smith, G.D. and Greally, J.M. (2016) 'Epigenome-wide association studies and the interpretation of disease-omics', *PLoS Genetics*, 12: e1006105. https://doi.org/10.1371/journal.pgen.1006105.

Biwer, C., Kawam, B., Chapelle, V. and Silvestre, F. (2020) 'The role of stochasticity in the origin of epigenetic variation in animal populations', *Integrative and Comparative Biology*, 60: 1544–57. https://doi.org/10.1093/icb/icaa047.

Bolt, I., Bunnik, E.M., Tromp, K., Pashayan, N., Widschwendter, M. and de Beaufort, I. (2020) 'Prevention in the age of personal responsibility: epigenetic risk-predictive screening for female cancers as a case study', *Journal of Medical Ethics*, 47(12): e46. https://doi.org/10.1136/medethics-2020-106146.

Brown, R.C.H. (2013) 'Moral responsibility for (un)healthy behaviour', *Journal of Medical Ethics*, 39: 695–8. https://doi.org/10.1136/medethics-2012-100774.

Buyx, A.M. (2008) 'Personal responsibility for health as a rationing criterion: why we don't like it and why maybe we should', *Journal of Medical Ethics*, 34: 871–4. https://doi.org/10.1136/jme.2007.024059.

Cavalli, G. and Heard, E. (2019) 'Advances in epigenetics link genetics to the environment and disease', *Nature*, 571: 489–99. https://doi.org/10.1038/s41586-019-1411-0.

Chadwick, R. and O'Connor, A. (2013) 'Epigenetics and personalized medicine: prospects and ethical issues', *Personalized Medicine*, 10: 463–71. https://doi.org/10.2217/pme.13.35.

Cheng-Guajardo, L. (2019) 'Responsibility unincorporated: corporate agency and moral responsibility', *The Philosophical Quarterly*, 69: 294–314. https://doi.org/10.1093/pq/pqy031.

Chiapperino, L. (2018) 'Epigenetics: ethics, politics, biosociality', *British Medical Bulletin*, 128(1): 49–60. https://doi.org/10.1093/bmb/ldy033.

Chiapperino, L. (2020) 'Luck and the responsibilities to protect one's epigenome', *Journal of Responsible Innovation*, 7(suppl 2): S86–106. https://doi.org/10.1080/23299460.2020.1842658.

Chiapperino, L. and Testa, G. (2016) 'The epigenomic self in personalised medicine: between responsibility and empowerment', in M. Meloni, S.J. Williams and P. Martin (eds), *Biosocial Matters: Rethinking the Sociology-Biology Relations in the Twenty-First Century*, Chichester: Wiley, p 616.

Chung, E., Cromby, J., Papadopoulos, D. and Tufarelli, C. (2016) 'Social epigenetics: a science of social science?', *The Sociological Review*, 64(suppl 1): 168–185. https://doi.org/10.1002/2059-7932.12019.

Cooney, C.A. (2007) 'Epigenetics – DNA-based mirror of our environment?', *Disease Markers*, 23: 121–37.

Dupras, C. and Ravitsky, V. (2016) 'The ambiguous nature of epigenetic responsibility', *Journal of Medical Ethics*, 42: 534–41. https://doi.org/10.1136/medethics-2015-103295.

Dupras, C., Saulnier, K.M. and Joly, Y. (2019) 'Epigenetics, ethics, law and society: a multidisciplinary review of descriptive, instrumental, dialectical and reflexive analyses', *Social Studies of Science*, 49: 785–810. https://doi.org/10.1177/0306312719866007.

European Commission Directorate-General for Health and Consumers (2013) *Addressing the New Challenges for Risk Assessment*, Luxembourg: Publications Office of the European Union.

Feinberg, J. (1968) 'Collective responsibility', *Journal of Philosophy*, 65: 674–88. https://doi.org/10.2307/2024543.

Fiorito, G., McCrory, C., Robinson, O., Carmeli, C., Rosales, C.O., Zhang, Y., et al (2019) 'Socioeconomic position, lifestyle habits and biomarkers of epigenetic aging: a multi-cohort analysis', *Aging*, 11: 2045–70. https://doi.org/10.18632/aging.101900.

French, P.A. (1984) *Collective and Corporate Responsibility*, New York: Columbia University Press.

Garcia-Reyero, N. and Murphy, C. (2018) *A Systems Biology Approach to Advancing Adverse Outcome Pathways for Risk Assessment*, New York: Springer.

Greco, J. (1995) 'A second paradox concerning responsibility and luck', *Metaphilosophy*, 26: 81–96. https://doi.org/10.1111/j.1467-9973.1995.tb00557.x.

Guerrero-Preston, R., Herbstman, J. and Goldman, L.R. (2011) 'Epigenomic biomonitors: global DNA hypomethylation as a biosimeter of life-long environmental exposures', *Epigenomics*, 3: 1–5. https://doi.org/10.2217/epi.10.77.

Hanna, N. (2014) 'Moral luck defended', *Noûs*, 48, 683–98. https://doi.org/10.1111/j.1468-0068.2012.00869.x.

Hedlund, M. (2012) 'Epigenetic responsibility', *Medicine Studies*, 3: 171–83. https://doi.org/10.1007/s12376-011-0072-6.

Heijmans, B.T. and Mill, J. (2012) 'Commentary: the seven plagues of epigenetic epidemiology', *International Journal of Epidemiology*, 41: 74–8. https://doi.org/10.1093/ije/dyr225.

Hens, K. (2017) 'Neurological diversity and epigenetic influences *in utero*. An ethical investigation of maternal responsibility towards the future child',, in K. Hens, D. Cutas and D. Horstkötter (eds), *Parental Responsibility in the Context of Neuroscience and Genetics*, Cham: Springer, pp 105–19.

Hooten, N.N. (2022) 'The accelerated aging phenotype: the role of race and social determinants of health on aging', *Ageing Research Reviews*, 73: 101536.

Huang, J.Y. and King, N.B. (2018) 'Epigenetics changes nothing: what a new scientific field does and does not mean for ethics and social justice', *Public Health Ethics*, 11: 69–81. https://doi.org/10.1093/phe/phx013.

Jensen, H. (1984) 'Morality and luck', *Philosophy*, 59: 323–30.

Jeremias, G., Gonçalves, F.J.M., Pereira, J.L. and Asselman, J. (2020) 'Prospects for incorporation of epigenetic biomarkers in human health and environmental risk assessment of chemicals', *Biological Reviews*, 95: 822–46. https://doi.org/10.1111/brv.12589.

Johannes, F. and Schmitz, R.J. (2019) 'Spontaneous epimutations in plants', *New Phytologist*, 221: 1253–9. https://doi.org/10.1111/nph.15434.

Jones, P.A. (2012) 'Functions of DNA methylation: islands, start sites, gene bodies and beyond', *Nature Reviews Genetics*, 13: 484–92. https://doi.org/10.1038/nrg3230.

Ladd-Acosta, C. and Fallin, M.D. (2019) 'DNA methylation signatures as biomarkers of prior environmental exposures', *Current Epidemiology Reports*, 6: 1–13. https://doi.org/10.1007/s40471-019-0178-z.

Lewis, H.D. (1948) 'Collective responsibility', *Philosophy*, 23: 3–18. https://doi.org/10.1017/S0031819100065943.

Loi, M., Savio, L.D. and Stupka, E. (2013) 'Social epigenetics and equality of opportunity', *Public Health Ethics*, 6: 142–53. https://doi.org/10.1093/phe/pht019.

MacIntyre, A.C. (2007) *After Virtue: A Study in Moral Theory* (3rd edn), Notre Dame, IN: University of Notre Dame Press.

MacKay, K. (2022) 'Public health virtue ethics', *Public Health Ethics*, 15: 1–10. https://doi.org/10.1093/phe/phab027.

Meloni, M. and Müller, R. (2018) 'Transgenerational epigenetic inheritance and social responsibility: perspectives from the social sciences', *Environmental Epigenetics*, 4(2): dvy019. https://doi.org/10.1093/eep/dvy019.

Mill, J. and Heijmans, B.T. (2013) 'From promises to practical strategies in epigenetic epidemiology', *Nature Reviews Genetics*, 14: 585–94. https://doi.org/10.1038/nrg3405.

Minkler, M. (1999) 'Personal responsibility for health? A review of the arguments and the evidence at century's end', *Health Education & Behavior*, 26: 121–41. https://doi.org/10.1177/109019819902600110.

Mitchell, K. (2018) 'Calibrating scientific skepticism – a wider look at the field of transgenerational epigenetics', *Wiring the Brain* [blog], 22 July. Available from: http://www.wiringthebrain.com/2018/07/calibrating-scientific-skepticism-wider.html [Accessed 14 May 2019].

Nagel, T. (1991) *Mortal Questions*, Cambridge: Cambridge University Press.

Panzeri, I. and Pospisilik, J.A. (2018) 'Epigenetic control of variation and stochasticity in metabolic disease', *Molecular Metabolism*, 14: 26–38. https://doi.org/10.1016/j.molmet.2018.05.010.

Peaston, A.E. and Whitelaw, E. (2006) 'Epigenetics and phenotypic variation in mammals', *Mammalian Genome*, 17: 365–74. https://doi.org/10.1007/s00335-005-0180-2.

Pentecost, M. and Meloni, M. (2018) 'The epigenetic imperative: responsibility for early intervention at the time of biological plasticity', *The American Journal of Bioethics*, 18: 60–2. https://doi.org/10.1080/15265161.2018.1526351.

Pettit, P. (2007) 'Responsibility incorporated', *Ethics*, 117: 171–201. https://doi.org/10.1086/510695.

Resnik, D.B. (2007) 'Responsibility for health: personal, social, and environmental', *Journal of Medical Ethics*, 33: 444–45. https://doi.org/10.1136/jme.2006.017574.

Rogers, W.A. (2004) 'Virtue ethics and public health: a practice-based analysis', *Monash Bioethics Review*, 23: 10–21. https://doi.org/10.1007/BF03351406.

Rothstein, M.A. (2013) 'Legal and ethical implications of epigenetics', in R.L. Jirtle and F.L. Tyson (eds), *Environmental Epigenomics in Health and Disease*, Berlin: Springer, pp 297–308.

Sand, M. (2018) *Futures, Visions, and Responsibility: An Ethics of Innovation*, Berlin: Springer.

Sand, M. and Klenk, M. (2021) 'Moral luck and unfair blame', *The Journal of Value Inquiry*, [online first] https://doi.org/10.1007/s10790-021-09856-4.

Santaló, J. and Berdasco, M. (2022) 'Ethical implications of epigenetics in the era of personalized medicine', *Clinical Epigenetics*, 14: 44. https://doi.org/10.1186/s13148-022-01263-1.

Sartorio, C. (2012) 'Resultant luck', *Philosophy and Phenomenological Research*, 84: 63–86.

Shanthikumar, S., Neeland, M.R., Maksimovic, J., Ranganathan, S.C. and Saffery, R. (2020) 'DNA methylation biomarkers of future health outcomes in children', *Molecular and Cellular Pediatrics* 7(1): 7. https://doi.org/10.1186/s40348-020-00099-0.

Smiley, M. (2022) 'Collective responsibility', in E.N. Zalta And U. Nodelman (eds), *The Stanford Encyclopedia of Philosophy* [online], 19 December. Available from: https://plato.stanford.edu/entries/collective-responsibility/ [Accessed 15 August 2023].

Statman, D. (ed) (1993) *Moral Luck*, Albany, NY: SUNY Press.

Szentkirályi, L. (2020) 'Luck has nothing to do with it: prevailing uncertainty and responsibilities of due care', *Ethics, Policy & Environment*, 23: 261–80. https://doi.org/10.1080/21550085.2020.1848185.

Talbert, M. (2019) 'Moral responsibility', in E.N. Zalta and U. Nodelman (eds), *The Stanford Encyclopedia of Philosophy* [online], 16 October. Available from: https://plato.stanford.edu/entries/moral-responsibility/ [Accessed 15 August 2023].

UK Health Security Agency (2023) Available from: https://research.ukhsa.gov.uk/about/ [Accessed 15 August 2023].

Velasquez, M.G. (1983) 'Why corporations are not morally responsible for anything they do', *Business and Professional Ethics Journal*, 2: 1–18. https://doi.org/10.5840/bpej19832349.

Vincent, N.A. (2011) 'A structured taxonomy of responsibility concepts', in, N.A. Vincent, I. van de Poel and J. van den Hoven (eds), *Moral Responsibility: Beyond Free Will and Determinism*, Dordrecht: Springer, pp 15–35.

Voigt, K. (2013) 'Appeals to individual responsibility for health: reconsidering the luck egalitarian perspective', *Cambridge Quarterly of Healthcare Ethics*, 22: 146–58. https://doi.org/10.1017/S0963180112000527.

Walker, M. (1991) 'Moral luck and the virtues of impure agency', *Metaphilosophy*, 22: 14–27.

Watson, G. (1996) 'Two faces of responsibility', *Philosophical Topics*, 24: 227–48.

Williams, B. (1982) *Moral Luck*, Cambridge: Cambridge University Press.

Wolf, S. (2001) 'The moral of moral luck', *Philosophic Exchange* 3(1): 1–17.

Zimmerman, M.J. (1987) 'Luck and moral responsibility', *Ethics*, 97: 374–86. https://doi.org/10.1086/292845.

Zimmerman, M.J. (2002) 'Taking luck seriously', *The Journal of Philosophy*, 99: 553–76. https://doi.org/10.2307/3655750.

4

Pictures at an Exhibition: Epigenetics, Harm and the Non–Identity Problem

Anna Smajdor

The non–identity problem

Many people share the intuition that a reproductive decision that results in the birth of a disabled child has harmed that child. Similarly, many people share the view that if it were possible to manipulate a person's genes to cure a disease, this would benefit that person. In turn, as we learn more about epigenetics, it has become apparent that switching genes on or off may also have effects that may be regarded as beneficial or harmful. Women who eat unhealthy diets may harm future offspring through epigenetic mechanisms affecting their eggs. Epigenetic alterations caused to the sperm of teenage boys who smoke may have adverse health effects on their future offspring. Seemingly, with epigenetics, the scope for prenatal, and even preconceptual, harm (or benefit) is vastly expanded.

In trying to work out how our choices harm offspring, we rely to some degree on assumptions about identity (Carlson et al, 2021). There are, of course, other accounts of harm, but I focus on comparative cases here. A comparative account of harm and benefit seems to work well in most everyday contexts. X harms Y by carrying out act A, if Y would have been better off had act A not occurred. However, it is difficult to make meaningful judgements about harm when the event that is deemed to be harmful is a necessary condition of someone's existence. This problem has been famously articulated by Derek Parfit as the non–identity problem (NIP). Suppose a woman – Anna – is receiving treatment for syphilis. She is also contemplating having a child. If she conceives now, while undergoing treatment, her baby – let us call it baby X – will be born with congenital syphilis. If she waits for six months until she is cured, any child she conceives will be free of syphilis. We may feel that she should wait and have a baby who will not suffer from syphilis. But if we try to explain this in terms of harm, we immediately flounder. The act A that causes baby X to have syphilis is also the act that brings baby X into existence. If Anna waits for six months, she does not benefit baby X. Instead, she will have a different child (baby Y, let us say), who will not have congenital syphilis. Thus, in this case, we lack the continuous component of identity that gives us grounds to say that this individual has been harmed, should Anna go ahead and conceive baby X before her treatment finishes.

Many bioethicists have interpreted Parfit's claims to indicate there is an important moral distinction between (1) reproductive choices that bring a new individual into existence, and (2) reproductive choices that harm (or benefit) an individual whose identity is already fixed. This distinction has been characterized in slightly different terminology by various contributors to the debate. In his paper on genome editing, Sparrow speaks of 'person-affecting' interventions versus 'identity-affecting' interventions. A person-affecting decision is the one that either benefits or harms the individual. Usually, Sparrow suggests, we think of gene therapy as being beneficial, thus it is 'person-affecting'. But the selection of one particular embryo over another is 'identity-affecting', and is thus not therapeutic (Sparrow, 2022). In contrast, in his analysis of whether mitochondrial replacement is therapeutic or not, Liao uses the terms 'qualitative identity' and 'numerical identity' (Liao, 2017). Alterations that affect qualitative identity are those that may harm or benefit an individual, while those that affect numerical identity are those that bring a different individual into existence. For the purpose of this chapter, I use the terms 'qualitative identity' and 'numerical identity' to capture this distinction. Interventions that bring about qualitative changes may be harmful or beneficial to the individual concerned. Interventions involving profound alteration to the individual may instead result in the creation of a numerically distinct identity.

The distinction between acts that affect qualitative versus numerical identity feeds into a particular understanding of moral responsibility in reproductive decisions. Ordinarily, we may think of ourselves as being morally responsible for the harm (or benefit) that we cause in our reproductive choices. Thus, if I carry out an intervention that results in offspring being blind or deaf, I am responsible for the harm this causes. But if there is no individual who has been harmed, it seems that this intervention does not easily fit into a harm-based moral framework. We neither harm nor benefit people by postponing reproduction, nor by selecting, for example, embryos that will result in deaf or blind offspring. I should clarify here that I take deafness and blindness as conditions that would be harmful to inflict on an individual: if you blind me, you have harmed me; if you deafen me, you have harmed me. This is not to say that the lives of deaf or blind people lack value, or are worth less than those of seeing or hearing people.

A further point of clarification here is that Parfit introduces another distinction in his analysis. That is, he suggests that an intervention that is identity-changing and brings about a disease or disability is not harmful unless it causes such terrible suffering that life itself becomes an intolerable burden for the individual concerned. Thus, in Parfit's original example, the woman who fails to wait until she is cured of syphilis has a different child from the one she'd have had if she'd conceived six months later. Her decision not to wait is a numerical identity-affecting choice. By implication, congenital syphilis is not the kind of condition that would trigger Parfit's harm threshold to be regarded as something harmful despite being numerical identity-changing (Parfit, 1984; Smajdor, 2014).

Some commentators – such as Wrigley, for example – claim that Parfit's argument is straightforwardly based on genetic essentialism (Wrigley, 2012). That is to say, the reason that baby X and baby Y are different is simply because they have different genes. But in many cases, the exact relationship between genes and identity is not fully unpicked. It is widely accepted that delaying pregnancy, and/or choosing between different embryos, sperm or eggs, are numerical identity-changing endeavours (Hope et al, 2012). In contrast, anything that happens to embryos after they have been 'created' is not likely to be identity-changing (Alonso et al, 2021). The broad acceptance of conception as the identity-fixing moment suggests that a genetic understanding of identity may be in operation here. Yet if genes are what fix identity, then it becomes difficult to show why genetic alteration at any age would be therapeutic rather than identity-changing.

In this chapter, I explore the non-identity problem against the background of findings in epigenetics. I suggest that there are some serious problems associated with a genetic account of identity. Some of these are resolved by incorporating epigenetics. But, in turn, an epigenetic account of identity raises new challenges to the idea of numerical identity through adult life. Crucially, these issues have an important bearing on the degree to which we are, or can be, morally responsible for benefitting or harming other people, whether in the peri-reproductive phase or afterwards.

What is genetic identity?

Humans go through many significant changes during their lifetimes. They change from blastocyst to fetus, baby, child, adult. They gain and lose weight. They may lose limbs, suffer strokes, give birth to children, gain and lose mental faculties. But we do not usually consider that these transformations call our identity into question. Perhaps this is because genes are indeed the essence of identity. My weight and height may change, I may donate a kidney or receive a heart, but my genetic make-up remains unchanged. Moreover, no other person could have emerged from the particular genes that form me. Therefore, it is tempting to think that what enables me to stay the same throughout the alterations that happen throughout my life must be the genes that lie, unchanged, at the very essence of my existence.

Parfit considers that identity is connected with the timing of one's conception. But how does this really play out in terms of identity? When the syphilis patient postpones having a child, Parfit assumes that this necessarily entails that the baby is born from a different egg and/or sperm. But is it the passage of time alone that is significant here, or the fact that the gametes, and hence the genes or chromosomes, are different? Parfit does not go into detail on this point, but some bioethicists assume, not unreasonably, that what matters is the gametes, or perhaps more specifically the genes or chromosomes, rather than the simple passing of time itself (Lewens, 2021). As Wrigley and colleagues state 'at the most fundamental level, genetic information has been seen as underpinning the numerical identity of persons' (Wrigley et al, 2015).

The view of genes or gametes as the essential component of identity makes sense, considering that we are often told that every individual human is unique. This uniqueness comes in part from the fact that we are formed from different gametes and each gamete is itself genetically unique. When a sperm or egg cell is formed, the genetic material is 'reshuffled', so that each gamete is different (this is why siblings who have the same parents are not identical). In theory, of course, it is possible that, by chance, this reshuffling might result in a replica of a previously produced gamete. To my knowledge, the implications of this have not been discussed anywhere in the literature. It is simply assumed that each gamete must be unique.

If each gamete is unique, it follows that each embryo will be genetically unique (except identical twins, and so on). From this, it is easy to reach the conclusion that my identity – what makes me *me* and no-one else – is my genes. Moreover, the language commonly used to discuss genes emphasizes both their uniqueness and their status as a code or blueprint. There is a tiny blueprint of me, replicated in every one of my cells. This idea of the code then opens the way for further metaphorical understandings of gene 'editing', of 'translation', of the genome as 'software' that directs the body's 'hardware'. The written material of the code is immutable, essential and necessary, whereas the physical manifestation is a mere edition, a product of the code. In this way, we can make sense of the fact that the physical materials that make us up are continuously replaced over time. Clearly a newly fertilized egg does not contain all the physical components that make the fully grown adult. What it does contain is the unique code that enables the body to transform physical matter into me. If we accept this way of looking at things, it seems clear that genes are indeed supremely important in our self-understanding. However, a focus on genes as the essence of identity leads to some problems. Concepts of genetic identity are confused, under-theorized and contradictory, yet they continue to feed into legislation and regulation, especially in the context of reproductive technologies (Ludlow, 2020).

Following the cloning of Dolly the sheep (Wadman, 2007), there was widespread concern about human reproductive cloning, and what it could mean for the identity of the cloned person. Bioethicists such as John Harris pointed out that we already have genetic clones in the world – namely identical twins – and yet we do not think the existence of such twins in some way calls into question whether either twin is really an individual (Harris, 1997). However, Harris and others do not always stipulate what kind of identity they take to be operational in the non-identity problem (Harris, 1993; Burley and Harris, 1999). In their discussion of *in vitro*-derived gametes, for example, Palacios-González and colleagues merely note that the decision to conceive with *in vitro*-derived gametes results in a 'completely different' child from the child conceived by other means (Palacios-González et al, 2014). Yet it is not clear in what sense this is necessarily true, and the authors do not explore the question further, although Williams and Harris (2014) do address the question of genetic identity in some depth in the context of the non-identity problem.

While many of those who have written about genetic identity mention in passing the challenge of accounting for identical twins, few people have actually gone to the effort of attempting to show why, if genes or gametes are the essence of numerical identity, twins are *not* the same person (Nordgren, 2008). It seems that, if we understand the identity component of Parfit's non-identity problem as being based on either genes or gametes, this is an unresolved issue. Indeed, perhaps it would be more accurate to say in this context that there is an identity problem, rather than a non-identity problem.

A further difficulty for a specifically genetic account of identity is that it implies that any intervention that alters the genetic make-up of an embryo, fetus, child or adult could also render that individual a different person. Sparrow appeals to the distinction between genetic interventions that would result in the creation of a numerically distinct individual and those that would benefit (or harm) an individual (Sparrow, 2021). He does not enter deeply into questions of genetic identity (see Gregg 2022, for example), but it does seem that genes play a crucial role, as he claims that 'genetic selection', including sperm sorting, 'involves determining which individual … comes into existence' (Sparrow, 2021). But if sperm sorting is identity-changing for essentially genetic reasons, it is not clear why genetic intervention undertaken in order to cure a disease is not also identity-changing (Räsänen and Smajdor, 2022). If this is the case, then such interventions cannot be therapeutic.

Equally counter-intuitive is the corresponding implication that any adverse impacts resulting from gene intervention (whether on embryos, children or adults) give us no moral grounds for concern. They cannot be said to harm the person, who simply becomes a genetically new individual. Thus, using the view of genetic identity implied by many of these writers, the harm principle seems not to apply. In one sense, genetic interventions are a free-for-all, devoid of risk; in another, they may properly be regarded as the province of the marketplace rather than medicine, as they can have no genuinely therapeutic impacts.[1]

Genetic essentialism

Genes are regarded as being important in understanding identity because they are the 'unit of transmission'. They have entered the popular imagination; almost every aspect of human behaviour has been attributed to 'a gene'. In the early days of the human genome project, people eagerly waited to learn how many thousands of genes a human being has. Being such a complex creature, the expectation was that there would be very many, probably around 100,000 (Salzburg, 2018). There was widespread shock, and even disappointment, when scientists announced that there were only around 30,000 genes after all (Salzburg, 2018). Tomatoes and grapes have more genes than we do. This being the case, it is clear that there cannot be a single gene that 'codes' for every human attribute.

Yet it seems that genes still dominate our understanding of identity and reproduction. The Warnock Report states that people experience 'a powerful urge to perpetuate their genes. … This desire cannot be assuaged by adoption' (Warnock, 1985). The UK's Human Fertilisation and Embryology Authority (HFEA) asserts: 'The wish for genetic offspring is a natural human aspiration' (HGA/HFEA, 1998). If we consider a characteristic to be 'in our genes', it is regarded as fixed in a way that other attributes are not. Even if we cannot say that there is one precise gene for every element of our personalities, we can surely say that the answer to the question of who we are is – at some level – in 'our genes'. This, after all, is the basis of genetic identity. It is commonly assumed that what makes me *me* and my children *mine* is genes. But what is it that is so special about genes?

Oftedal notes the way in which language and metaphor influence our understanding in this context. Metaphors can be partially accurate and represent one aspect of a phenomenon while failing to capture other aspects. In using certain metaphors, we may reinforce dominant tropes, or prioritize certain understandings over others. Or the metaphors may turn out simply to be inaccurate or misguided (Oftedal, 2022). The language of genes has a power that is both metaphorical and/or symbolic. Genes represent the new language of kinship and connection. While people in the past spoke of 'blood', modern day people appeal to genes. Genes both unite and divide: they give us identity, and bind us to each other (and differentiate us) in specific ways. But genes are of course more than mere symbols. Perhaps we can gain a better understanding of how they interact with identity by looking more closely at what they are, and how they behave.

Genes have a particular function: coding for proteins. These proteins then have a bearing on our looks, behaviour, health and preferences. But the relationship between genes and identity is a little more complex than this. This is because not all of the genes that we have are expressed. Some of what makes us what we are depends on exactly which genes are operative, so to speak. This is where epigenetics comes in. Of all the genes we have, some are switched on, some are switched off, and others are active only to a degree. This switching process is, crudely speaking, epigenetics. Environmental influences, both inside and outside the body, interact with our genes and trigger this process. This doesn't change what genes we have, but it does change which genes are active, and to what degree. This raises the question of whether we should understand genetic identity in terms of genes themselves, or in terms of gene expression, that is epigenetics.

Epigenetics

It seems clear that epigenetic alterations can be extremely significant. Raz et al (2019) point out that what makes a bee a queen bee is not determined by genes per se, but by epigenetics. All bees have the potential to be queens, but only bees in whom the relevant epigenetic processes occur will ultimately achieve queendom. Their point here is not to say that epigenetics holds the key to similar transformations in human beings, but rather that things that are not genetically determined can have a profoundly significant effect on one's physical appearance and capabilities, perhaps to the extent that they can be deemed numerical identity-changing.

Although epigenetics has not captured the public imagination to the same extent that genes themselves did, it is increasingly evident that environmental factors may affect gene expression in ways that are heritable. Thus, the 15-year-old boy who smokes may go on to father children who are more likely to become obese (Liu et al, 2022). The woman who enjoys her stressful job may be unaware that her eggs are being affected epigenetically by that stress, in ways that will have health consequences for her children or even grandchildren (Turkmendal and Liaw, 2022). Should we regard these as cases of harm to future offspring? If so, how far, if at all, should we feel morally responsible for such harm? Scholars have noted that, as knowledge of epigenetics filters through into the public domain, it may play into existing prejudices and assumptions, for example, that women are responsible for the health of their future offspring to a greater extent than men (Lappe, 2016; Dubois et al, 2019).

Our understanding of epigenetics is still relatively limited. But, as it expands, its significance and moral implications become increasingly pressing. As a field, epigenetics is still fluid and contested; the discourse of epigenetics is in the process of being 'encoded' by scientists and 'decoded' by the public (Raz et al, 2019). Yet it seems that this process of decoding is still very much a work in progress, at least insofar as it relates to ideas about genetic identity.

The genome as an art gallery

One way of conceptualizing identity, encompassing the complexity of epigenetics, is to think of the whole person's genome as an art gallery. To be a human is to have a gallery for the display of one's personal exhibition. All humans have this in common, as well as some similarities perhaps in layout and classification (for example, we might agree that there are a certain number of floors, that there are specific areas set aside for painting, drawing, sculpture, conceptual art, and so on). These correspond with fixed channels within which genetic variation may occur. Human development follows a certain fixed sequence. We have certain characteristics in common: eyes, ears, arms, legs, and so on. But, within these fairly rigid confines, there is enormous scope for variation, not just in terms of what genes one has, but what genes are expressed, or 'on display'. The items that may or may not be on display at a certain time are the genes.

Art galleries almost invariably have large amounts of stock, in addition to what is on view to the public at any time. Only a small proportion of the stock is on display, and the exact make-up of the exhibitions will change over time. The items in the exhibition are thus crucial, in this analogy, to making us who we are. What I see when I look at you is some of the pictures that are on display: your exhibition. But this is only a fraction of what could have been displayed. In this analogy, it is chromosomal inheritance that determines what artworks (genes) are in the building, and at least partly epigenetics that determines what is actually on display. The active genes are the paintings in the exhibition. The inactive ones – those that have been silenced – are in the stock room, and some of them may never emerge during the lifetime of this gallery.

So should we think of a person's identity as being the entire collection, most of which will never be shown, or the material on display, which may change over time? There are three possible scenarios here:

1. Items on display are periodically switched with those in the stock room. This means that, at different times, there are different exhibitions on display, perhaps with no overlap between pictures. These are the equivalent of epigenetic changes, which happen continuously throughout a normal human life (although not usually through any deliberate intervention). Would this change a person's identity?

2. We can switch some of the pictures in the stock room for those from other collections. This would involve significant changes to the potential for displaying different exhibitions. But as long as the pictures switched are those that were in the stock room, there would not necessarily be any difference in what is in fact on display. Visitors to the gallery would never know that these changes have been made. This is the equivalent of gene alterations that are never expressed.
3. We can display some of the pictures obtained from other collections, so that not only do we now have pictures from other collections within our gallery, as in scenario 2 above, but these are also on show, meaning that the difference can be seen and observed by visitors to the gallery. This is the equivalent of gene alterations that are expressed. Which, if any, of these scenarios would change a person's identity?

Scenario 1 involves some significant differences in what is perceived by visitors to the gallery. Or, to return to the question of identity, changing which genes are 'switched on' may make profoundly striking alterations to a person's appearance and behaviour. In theory, such differences may make it hard to recognize that this is indeed 'the same' person. Yet, in purely genetic terms, nothing has changed. The same genetic complement remains intact.

Scenario 2 seems challenging in the context of identity questions. The transfer or alteration of genes – if they are never expressed – may have little relevance in terms of changes to the outward manifestation of our identity, nor will it have a bearing on the subjective experience of the person in question. But if we decide that scenario 2 is not identity-changing, this seems to imply that what really matters is gene expression. If so, it is not so clear that there is a distinction to be made between scenario 1 and scenario 3. That is to say, any epigenetic change, whether brought about through epigenetic factors, or caused by gene replacement plus epigenetic factors, seems to be identity-changing.

However, it is not so clear that this gives us a good mandate for choosing genes as the basis for personal identity. If we say that epigenetic changes as in scenario 1 are not numerical identity-changing, any epigenetic alteration is open to question in terms of whether it causes harm. It may be harmful (or indeed beneficial). This may be true, whether it happens at the gametic, embryonic or adult stage. If this is the case, any time someone suffers from a disease caused by epigenetic changes associated with pollution, for example, they have been harmed.

But if we take it that the genes themselves, whether in the stock room or on display, are what is important here, it suggests that, if we undergo a process that changes a gene, that goes beyond mere switching on and off of existing genes, for example CRISPR, or genetic modification, we change someone's identity. A new person comes into being. We cannot be said to harm them, then, even if the intervention turns out to be pretty terrible.

When we think of identical twins, clearly they are genetically identical – they share exactly the same genes. To continue the gallery analogy, they have exactly the same paintings in stock. But it seems absurd to say that identical twins are in fact numerically the same person. Evidently, they are not! And, if not, then it seems that numerical identity cannot be determined purely by genes.

Perhaps we can make some sense of this by noting that even identical twins are not in fact identical to each other from the moment of birth to death. They become increasingly different epigenetically as life goes on. Different genes are expressed throughout the life course, partly as a result of 'programming' (a newly fertilized egg will go through a number of epigenetic changes as it transitions from gamete to embryo, to fetus, and so on), and partly through contextual and environmental interactions (Flintoft, 2005). Of course, everyone knows that identical twins don't always look identical. They may dress differently. One may become overweight, the other muscly. One who lives in a polluted area may develop respiratory problems; one who lives in a sunny place may develop skin cancer. But we would normally regard these as being environmental differences.

However, once we understand that these effects are in fact epigenetic, it becomes clear that we can no longer rely on the simple genetic/environment, biological/social or nature/nurture dichotomies that used to offer a convenient way of distinguishing between phenomena that were given and those that are merely contingent. The 'givenness' of genetic identity thus starts to look less plausible. And, in this case, it is not so clear that numerical genetic identity as a core component of the non-identity problem can really sustain the weight it is expected to bear.

We can ask the question like this: should we assume that the genetic inheritance of identical twins is what constitutes identity? In which case, we seem to have to say that numerically they are the same person. Or should we say that identity is related in some way to gene expression instead of (or as well as) the genes themselves? If so, going back to the gallery analogy, we can say that the twins start out with identical stock and identical items in the exhibition rooms, but that the exhibitions become increasingly different over time. If we are looking for a way of making sense of the twins as unique individuals, then it seems that epigenetics gives us that possibility.

But if we go with the idea that gene expression, rather than genes themselves, is what is crucial for identity, this raises a host of other problems. For a start, as suggested, the epigenome is fluid throughout a person's lifespan, and even before. If we pinpoint the epigenome at the moment of conception as being the thing that counts, it seems that we are committed to the idea that every subsequent epigenetic change in fact creates a new person. If this is the case, then anything that causes epigenetic changes should be of no concern because they merely create different people. If I am poisoned by air pollution, or experience a period of starvation, or am bullied to the extent that the stress causes epigenetic changes – none of these things actually harm me in the sense of making my life worse, they merely cause a different individual to come into being. If we take harm to be necessarily connected with making someone's life worse than it would otherwise have been, and if epigenetic changes are identity-changing, they may cause me to cease to exist, but they do not harm me. (There is a wealth of literature on the question of whether death is a harm to the individual who dies, but, for the purpose of this discussion, I do not attempt to enter into that debate here.)

Thus, as I have shown, a focus on genes seems to be problematic in terms of understanding identity in a way that makes sense of our intuitions about harm and benefit. This remains true even if we take epigenetics into account. However, perhaps greater clarity can be achieved by looking more closely at gametes, rather than attempting to understand identity in a way that divorces (epi)genetics from the biological context of the cell.

Eggs and sperm

There are many ways in which people change through their lives without necessarily regarding their identities as having changed. But each person is the product of only one set of gametes. Perhaps it is here that the answer lies. However, again, technology raises new questions. We can now switch some components of an egg cell, so that the mitochondrial DNA comes from one woman and the nuclear DNA from another. This raises the question of whether such a switch is identity-changing. In their discussion of this, Wrigley et al (2015) argue that there is a difference between two types of mitochondrial DNA transfer. In one case, there is a delay between the creation of the modified egg, and its fertilization, meaning that the sperm that would have fertilized the unmodified egg is no longer available. In short, a different sperm is involved in each case, thus creating a different individual. In their analysis, they move seamlessly between discussing gametes and genes, as though there is nothing important about a gamete except its genes. And genes, in this context, are taken to be chromosomal genes contained within the nucleus, rather than mitochondrial genes.

Lewens objects to this manoeuvre as there is no guarantee that the same sperm that would have fertilized the egg 'naturally' will be the same one to fertilize it when either of the two techniques are used (Lewens, 2021). He goes further, suggesting that Wrigley and colleagues place too much emphasis on genes in their interpretation of Parfit. They take it that Parfit is committed to gametic essentialism. For Lewens, the important thing in Parfit's account is time dependency. A child conceived today cannot be the same as a child conceived at a later point in time. So is time the key factor for identity? We can explore this further by considering what Parfit's argument would look like if the same gametes were used, but at different periods of time. Suppose, then, I am considering having a child now. I have the egg and the sperm cell that I propose to use, right in front of me. I could go ahead now. Alternatively, I could decide to freeze the gametes, and thaw them at some future point, perhaps six months, or even years. In this way, the time question can be addressed independently of the gamete question. Is it the same child in both cases?

To revert to Parfit's example, we can no longer assume that a woman who is deliberating whether to have a baby now or in six months will necessarily have a baby conceived with different gametes. (At the time Parfit was writing, these possibilities were not available, nor was it expected that egg freezing, in particular, would ever be a viable option.) Similarly, Wrigley et al (2015) are wrong to insist that the time difference between the two types of mitochondrial transfer has necessary implications for the sperm that ultimately fertilizes the egg. The link between time dependency and gamete variability no longer holds.

This possibility enables us to unpick the connections between genes, epigenetics and gametes to some degree. Genetically, the child conceived today in my example, is identical with the child conceived in six months or six years, using the same egg and sperm. But if we consider this from the epigenetic perspective, there may indeed be differences between Baby A and Baby B. The freezing process – and perhaps the passing of time itself – results in epigenetic changes to the egg, even though these are not deliberately brought about (Barberet et al, 2020). The culture medium, the temperature, the containers in which they are kept, temperature variations and light will all have an effect on the gene expression of the gametes. These effects may be minuscule or profoundly significant in terms of the offspring's appearance, health or even survival (Powledge, 2011). Based on this, it might make sense to say that babies A and B are not the same individual. If so, this reinforces the idea that epigenetics is involved in identity, at least to some degree. However, if we accept this, it also seems to suggest that whatever we do to the egg, sperm or embryo that results in these epigenetic changes, cannot be construed as harmful. Once again, we are looking at scenarios that give rise to new individuals, rather than affecting existing individuals.

We can make this even more explicit by considering a few different scenarios. Suppose an egg cell and a sperm cell are harvested. Each gamete is currently being stored in a laboratory and the plan is to use them to create an embryo.

Scenario 1: The sperm is injected into an egg, resulting in an embryo.

Scenario 2: The egg is deliberately placed in an acid bath that brings about certain epigenetic events, including activation of a particular gene. The sperm is injected into the adapted egg, resulting in an embryo. Individuals in whom this gene is activated will be blind.

Scenario 3: The sperm is deliberately placed in an acid bath that brings about certain epigenetic events, namely activation of a particular gene. The adapted sperm is injected into the egg, resulting in an embryo. Because of the intervention in the sperm, the offspring will be blind.

In all of the scenarios, the same gametes are involved, and hence the same genes. But can we say it is the 'same' baby in each case? If so, have we harmed it by undertaking the interventions described in scenarios 2 and 3? According to the basic Parfitian account, it seems that these babies are necessarily the same child. Therefore, the baby in scenario 2 has indeed been harmed (if we regard blindness as being harmful). Although its genetic make-up has not changed, it has undergone something that makes its life worse.

But how can it harm a person to put an egg in an acid bath? An egg is not a person. If an egg has an identity, it is certainly not the same sort of identity that an embryo has. It only makes sense for us to say the child is harmed in scenario 2 if we somehow regard the destiny of the egg as being fixed: this egg is inexorably fated to be fertilized by this particular sperm. Yet, in practice, the destinies of eggs and sperms are not fixed in this way. The sperm injection scenario is one technique that makes it less contingent which sperm meets which egg. But this is only one of many contingencies that can occur in the 'lifetime' of a gamete. Any sperm may meet any number of different eggs. The vast majority of sperm, and most eggs, will not result in any conception at all. Any gamete may undergo epigenetic changes as a result of the passing of time or environmental circumstances. The nature of the epigenetic change brought about by the decision to put the egg in the acid bath in scenario 2 is no more contingent than our determination that the egg should be fertilized by this particular sperm rather than another, or the decision to have a baby now rather than waiting six months. In all these cases, our choices affect the outcomes. But admitting that our choices affect outcomes is not the same as admitting that our choices harm individuals.

Because of this, the way that gametes per se are connected with identity seems deeply complicated, whether we are focusing on genes, epigenetics, or both. Before an embryo is formed, there are two separate cells that may or may not come together. Thinking about events that may happen after this is akin to regarding a person as being a potential spouse. If they marry, their past choices and behaviour will indeed have impacts on their marriage that will make it go more or less well. But it would seem both excessively demanding and metaphysically implausible to suggest that the potential 'marriage' is an entity that requires moral consideration before it comes into existence. The concept of harming a marriage is of course metaphysically problematic in a way that harming a person is not. But harming a future person who may come into being through a series of events pertaining to a particular gamete seems just as contentious.

Of course, the moral implications of harming a future spouse, or a future person, may seem different depending on the intentions of the moral agent. Suppose I do intend to marry, I am already engaged to a specific person with whom I plan to spend my married life. And I embark deliberately on a course of action that will foreseeably make this marriage go worse than it would have done otherwise – without any countervailing reasons. Or suppose I do intend that this particular egg that I have harvested will be fertilized by a particular sperm that I have identified, and I decide to put the egg in an acid bath that will affect gene expression in a way that makes the resulting offspring blind. In both these cases, it seems clear that I, the moral agent, have behaved unethically. The more certain I am that the marriage or conception will take place, the more blameworthy my actions seem. But this unethical behaviour cannot easily be reduced to harm to another individual; indeed, it does not fit into a consequentialist framework at all. In both cases, the moral opprobrium may be fairly easily explained with reference to virtue ethics or deontology. But, of course, the non–identity problem does not concern itself with these moral theories.

Thus far, then, I suggest that we cannot associate gametes with identity on the basis of their genetic and/or epigenetic connection with future individuals. Part of the reason for this, as I have shown, arises from the tenuous link between a gamete and the future individual whose existence it may facilitate. This gives rise to a counter–intuitive interim conclusion that it is not harmful to place an egg in an acid bath in order to bring about the blindness of a future individual. The counter–intuitive nature of this conclusion may be powerful enough to push us back towards a claim that there is a relevant kind of identity connecting the gamete and the resulting offspring.

So far in this section, I have focused on gametes. Part of my argument against being able to harm offspring via (epi)genetic interventions is both the problem of (epi)genetic identity – if we change the identity, we can no longer be said to harm the individual – and the problem that the individual who is to be harmed is remote from the gamete in ways that make it problematic to regard them as the same individual, despite the fact that the resulting person has the genes that that gamete conveys.

It seems that, whichever way we look, we run into counter–intuitive conclusions. If identity is genetic, then gene therapy is not therapeutic at all, and identical twins are numerically the same person. If identity is epigenetic, we have no stable identity throughout our life course, but are constantly changing. Moreover, environmental stresses and pollution that bring about disease-causing epigenetic changes cannot be said to 'harm' those who are affected by them.

However, so far, we have been looking at gametes only in the context of the genes that they convey and/or the role played by epigenetic factors. But could there be some other element of identity that pertains to gametes other than their (epi)genetic make-up? In order to answer this, it is worth looking a bit more closely at what gametes actually are.

The egg: first among equals?

Lewens (2021) has argued that many of the discussions of the non-identity problem, including some of those centring on mitochondrial DNA replacement, are founded on misapprehensions about the role of genetic identity in Parfit's discussion. He explores the idea that perhaps it is the egg that is determinative of identity. The egg is far larger than the sperm, and also contributes mitochondria to the offspring, which the sperm does not. So there are some grounds for saying that, although the chromosomal contribution is equal, we should regard the egg as being 'special'. If so, that may give us reason to claim that, when we place the egg in the acid bath, it is harmful (or beneficial) (Lewens, 2021). In contrast (by implication), the question of exactly which sperm fertilizes the egg is not central to the identity of any resulting embryo, or baby. This fits neatly with the facts pertaining to gametes: women generally release one egg per month, while men produce millions of sperm with each ejaculation. Sperm are designed to be 'disposable' in a way that eggs are not. This month's egg, without medical intervention, will retain the same features, whenever it is fertilized. But even just one ejaculation gives us myriad possibilities as to which sperm might ultimately feature in the eventual offspring.

An egg-focused account of identity has some appeal to it. It moves away from the reductive focus on genes. Gametes, after all, are cells with a specific kind of function. Cells are often described purely in terms of their genetic contents. The nucleus is where our 46 chromosomes are found, and the chromosomes in turn contain the genes, which are switched on or off by epigenetic factors. Thus, traditionally, the nucleus is the focus of interest for those who are concerned with genes. Yet a cell is vastly more than a repository of genes. It is a highly complex organism, delicately structured around specific functions, with a predictable biological trajectory, much of which has little direct bearing on genes per se (Alberts, 2017).

Liao (2017) argues for a particular understanding of cellular continuity, whereby changes to the organism's structure or functions may change its identity. It is this that makes it possible to say that the egg is no longer an egg after fertilization. It has changed identity and become something else. Based on this view, we can change an egg's numerical identity, Liao says, by, for example, changing some core feature of the cell, such as replacing the mitochondria. To replace the mitochondria does not constitute a 'therapeutic' intervention for the egg on Liao's account, but rather creates a new, reconstructed egg. However, this is not for primarily genetic reasons, but because a significant component of the egg has been altered. Accordingly, perhaps one could argue that, on the basis of cellular continuity, epigenetic alterations – if sufficiently sweeping – would also affect the identity of the egg. However, it is not clear in fact that epigenetic changes would fulfil Liao's criteria for a change in cellular continuity, and hence identity. After all, eggs do ordinarily undergo epigenetic changes, as do all cells.

A final point here is that, while the egg clearly has properties that the sperm lacks, and these may lend weight to the idea that the egg is what matters for identity, and thus for questions of harm, this has further counter-intuitive implications. In the acid bath scenarios above, scenarios 2 and 3 involve an egg and sperm, respectively, being subjected to alterations that result in blindness in the offspring. Should our conclusions as to whether the intervention is numerical identity-changing, or qualitative identity-changing – and thus harmful –differ depending on whether it is the sperm or egg that is subjected to the acid bath? Given that the outcome for the offspring is the same in each case, to insist that there is a difference seems to push 'egg exceptionalism' too far. This is equally the case whether genes or epigenetics are the key focus.

Epigenetics and the fluidity of biological identity

In this chapter, I have explored Parfit's non-identity problem from the perspective of new developments in epigenetics. I discuss the significance of the distinction between numerical and qualitative identity-changing interventions in reproductive ethics, and show how this distinction has become crucial in establishing whether reproductive decisions harm offspring. I analyse the problems that are inherent in a primarily genetic understanding of identity in the context of Parfit's work. I consider the question of whether an approach that encompasses epigenetics is preferable, and show that, although it resolves some problems, it raises others. To focus on genes as the essence of numerical identity seems implausible from a biological perspective: we would be forced to regard identical twins as one person. Moreover, it is not possible to explain how gene alteration could be therapeutic, rather than simply causing the existence of a new person. From the epigenetic perspective, we can recognize genetically identical twins as two distinct individuals, yet, if we assume that epigenetics is the key factor in numerical identity, it suggests that we are many different people throughout our lives. In turn, this suggests that I cannot be harmed or benefitted by interventions or events that cause epigenetic changes. Instead of making me better or worse off, the intervention simply makes me cease to exist.

Bioethicists often pride themselves on their ability to keep pace with scientific developments, to move forward with science, rather than trail behind it. But perhaps we are misguided in seeking answers to our moral questions by looking closely into the biology of human beings. Part of what compels us to do this is the utilitarian bent, shared by a number of bioethicists, that requires us to calculate risks and harms, and to apply these calculations to discrete, identifiable, individuals. If we set aside the question of harm for a moment, we can look at the complexity of biology and its implications for metaphysics and philosophy, independently of our moral preoccupations. Boniolo and Testa, for example, in discussing the identity of living things, state that 'any living being is the result of the epigenetic processes that have regulated the expression of its genome' (Boniolo and Testa, 2012, p 279). Continuity, on their view, lies in the smoothness of the epigenetic processes, one leading into the next. They note the difficulty in accounting for individuality and identity historically, and point out that genetics has come to play a reductive function in this.

Identity has, for bioethicists, commonly been addressed independently of environment; biology and society are separate – one is given, the other contingent. It is this that has enabled the discourse of genetic essentialism to gain momentum, and to support a separation between fixed biological identity and environmental influences/impacts that may cause harm to a single stable self. Epigenetics makes these distinctions crumble away. Boniolo and Testa (2012) observe that the environment for a gene is the cell in which it resides. For a cell, the environment is the other cells that surround it. Viewed in this way, the distinction between genes and environment is highly implausible; genes are environment because they are part of cells.

The blueprint metaphor for genes is wrong, or perhaps more accurately, it is only a very partial account. Cells from the body, such as heart, lung, brain or muscle cells, do not contain within themselves all that is needed for their development and successful functioning. Rather, they depend on the 'external microenvironment' in which find themselves (Odorico et al, 2001). A heart cell that is isolated and cultured will not follow the same trajectory as its counterparts within a functioning heart, as it does not 'recognize' its environment, and cannot respond to the stimuli provided by other cells and structures. The point here is that biological boundaries are more fluid than is commonly imagined. This being the case, the discreteness of entities, whether genes, chromosomes, cells, organisms or species, is not always conceptually sustainable. Nor is the biological necessarily distinguishable from the social or contextual – even at the cellular level.

Boniolo and Testa state that:

a living being, in any instant of its life, is nothing but the result of all the epigenetic processes that, in the course of time, have (linearly or non-linearly) causally molded all of its interrelated phenotypic modules (be they metabolic, immunological, nervous, behavioral, and so on); i.e., in any instant of its life a living being is its whole phenotype intended as the outcome of its epigenetic history. (Boniolo and Testa, 2012, p 287)

Perhaps here, Boniolo and Testa push a little too far: 'nothing but' has a reductive ring to it. Nevertheless, their account seems plausible. The smoothness of biological transitions enables us to perceive identity. Epigenetics is necessarily a part of this, if not the whole. When sweeping changes or abrupt alterations are made, our convictions concerning identity are shaken. Interestingly, this is often how things work in normal life. A person who loses a limb, experiences profound illness, gives birth to a child or suffers a stroke may be regarded by themselves and others as becoming a 'different person'. Ordinarily we might dismiss this as whimsical thinking, but perhaps it is closer to the truth than we realize.

To conclude, biology gives us fascinating insights into the questions that have always occupied philosophers. Epigenetics gives us new ways of understanding the interactions between genes and environment. It also gives rise to new moral concepts such as 'epigenetic harm'. In turn, this seems to lead towards new possibilities for ascribing responsibility: epigenetic responsibility. Yet epigenetics does not offer easy answers to any of these moral questions. In some respects, it seems that we simply re-clothe existing moral concerns and obligations in the new language of epigenetics. I suggest that we do not really need to couch the arguments against environmental pollution, smoking, heavy drinking, excessive stress or putting gametes in acid baths in terms of harm to specific individuals. There are moral reasons to avoid these activities (all other things being equal) without needing to carve the world up in precisely the way that Parfit might have thought necessary, and that continues to influence discussions and policies in reproductive ethics.

Note

[1] Medicine is widely regarded as having a special obligation of beneficence that sets it apart from other endeavours. I do not necessarily endorse this idea, but it seems uncontentious to say that interventions that are not, and cannot be therapeutic in any medical sense, are not obviously the domain of medicine.

References

Alberts, B. (2017) *Molecular Biology of the Cell*, New York: Garland Science.

Alonso, M. and Savulescu, J. (2021) 'He Jiankui's gene-editing experiment and the non-identity problem', *Bioethics*, 35(6): 563–73.

Barberet, J., Barry, F., Choux, C., Guilleman, M., Karoui, S., Simonot, R., et al (2020) 'What impact does oocyte vitrification have on epigenetics and gene expression?', *Clinical Epigenetics*, 12(1): 121.

Boniolo, G. and Testa, G. (2012) 'The identity of living beings, epigenetics, and the modesty of philosophy', *Erkenntnis*, 76(2): 279–98.

Burley, J. and Harris, J. (1999) 'Human cloning and child welfare', *Journal of Medical Ethics*, 25(2): 108–13.

Carlson, E., Johansson, J. and Risberg, O. (2021) 'Well-being counterfactualist accounts of harm and benefit', *Australasian Journal of Philosophy*, 99(1): 164–74.

Dubois, M., Louvel, S., Le Goff, A., Guaspare, C. and Allard, P. (2019) 'Epigenetics in the public sphere: interdisciplinary perspectives', *Environmental Epigenetics*, 5(4): dvz019.

Flintoft, L. (2005) 'Identical twins: epigenetics makes the difference', *Nature Reviews Genetics*, 6(9): 667.

Gregg, B. (2022) 'The person-affecting/identity-affecting distinction between forms of human germline genome editing is useless in practical ethics', *The American Journal of Bioethics*, 22(9): 49–51.

Harris, J. (1993) 'Is gene therapy a form of eugenics?', *Bioethics*, 7(2–3): 178–87.

Harris, J. (1997) '"Goodbye Dolly?" The ethics of human cloning', *Journal of Medical Ethics*, 23(6): 353–60.

Hope, T. and McMillan, J. (2012) 'Physicians' duties and the non-identity problem', *The American Journal of Bioethics*, 12(8): 21–9.

Human Genome Authority/Human Fertilisation and Embryology Authority (HGA/HFEA) (1998) Cloning Issues in Reproduction, Science and Medicine, London: Human Genetics Advisory Committee, December, para 4.6.

Lappé, M. (2016) 'Epigenetics, media coverage, and parent responsibilities in the post-genomic era', *Current Genetics Medicine Report*, 4: 92–7. https://doi.org/10.1007/s40142-016-0092-3.

Lewens, T. (2021) 'The fragility of origin essentialism: where mitochondrial 'replacement' meets the non-identity problem', *Bioethics*, 35(7): 615–22.

Liao, S.M. (2017) 'Do mitochondrial replacement techniques affect qualitative or numerical identity?', *Bioethics*, 31(1): 20–6.

Liu, Y., Chen, S., Pang, D., Zhou, J., Xu, X., Yang, S., et al (2022) 'Effects of paternal exposure to cigarette smoke on sperm DNA methylation and long-term metabolic syndrome in offspring', *Epigenetics & Chromatin*, 15(1): 3.

Ludlow, K. (2020) 'Genetic identity concerns in the regulation of novel reproductive technologies', *Journal of Law and the Biosciences*, 7(1): lsaa004.

Nordgren, A. (2008) 'Genetics and identity', *Public Health Genomics*, 11(5): 252–66.

Odorico, J.S., Kaufman, D.S. and Thomson, J.A. (2001) 'Multilineage from human embryonic stem cell lines', *Stem Cells*, 19: 193–204.

Oftedal, G. (2022) 'The metaphorical role of the histone code', in S. Wuppuluri and A.C. Grayling (eds), *Metaphors and Analogies in Sciences and Humanities*, Cham, Switzerland: Springer, pp 253–67.

Palacios-González, C., Harris. J. and Testa, G. (2014) 'Multiplex parenting: IVG and the generations to come', *Journal of Medical Ethics*, 40(11): 752–8.

Parfit, D. (1984) *Reasons and Persons*, Oxford: Clarendon Press.

Powledge, T.M. (2011) 'Behavioral epigenetics: how nurture shapes nature', *BioScience*, 61(8): 588–92.

Räsänen, J. and Smajdor, A. (2022) 'Epigenetics, harm, and identity', *The American Journal of Bioethics*, 22(9): 40–2.

Salzberg, S.L. (2018) 'Open questions: how many genes do we have?', *BMC Biology*, 16(1): 94.

Smajdor, A. (2014) 'How useful is the concept of the 'harm threshold' in reproductive ethics and law?', *Theoretical Medicine and Bioethics*, 35(5): 321–36.

Sparrow, R. (2021) 'Human germline genome editing: on the nature of our reasons to genome edit', *The American Journal of Bioethics*, Apr 16: 1–2.

Turkmendag, I. and Liaw, Y.Q. (2022) 'Maternal epigenetic responsibility: what can we learn from the pandemic?' *Medicine, Health Care and Philosophy*, 25: 483–94.

Wadman, M. (2007) 'Cloning special: Dolly – a decade on', *Nature*, 22;445(7130): 800.

Warnock, M.A. (1985) *Question of Life*, Oxford: Blackwell.

Williams, N.J. and Harris, J. (2014) 'What is the harm in harmful conception? On threshold harms in non–identity cases', *Theoretical Medicine and Bioethics*, 35(5): 337–51.

Wrigley, A. (2012) 'Harm to future persons: non-identity problems and counterpart solutions', *Ethical Theory and Moral Practice*, 15(2): 175–90.

Wrigley, A., Wilkinson, S. and Appleby, J. (2015) 'Mitochondrial replacement: ethics and identity', *Bioethics*, 29: 631–8.

5

Epigenetics, Parenthood and Responsibility for Children

Daniela Cutas

Those who contribute biologically to a child´s identity are commonly seen as that child's biological parents. While people who raise a child will of course have a great impact on the child's potential to flourish, they will not do so as biological contributors: they can only contribute in other ways. This divide between biological and social contributors to children's lives has often been taken for granted in discussions of parenthood, responsibility for children and the meaning of biology in determining relationships with children. For those who place great value on the biological, or, more specifically, on the genetic link between parents and children, the divide is clear and meaningful. Different people may form parent-like relationships with children, but the question of who our biological parents are has a clear answer: they are those who we are made of – those who determine our biology – and our biology is determined by our genetics. Moreover, people who create children biologically are often seen prima facie as the children's parents and holders of a special kind of responsibility in relation to them. Children are made by their biological parents, and their very biology depends on that of their biological parents.

That straightforward determination is being challenged by an increasingly profound understanding of the inter-dependent relationship between genes and the environment. Findings in epigenetics suggest that the environment in which a child is raised influences which of their genes are expressed and how, in ways that seem to be heritable. In this way, epigenetics blurs the boundary between genetics and the environment, and thus allows an analysis of contributions to children's lives that goes beyond classical dualistic categories such as genetic versus environmental or biological versus social. It is this analysis that I plan to undertake in this chapter, against the background of the attribution of parenthood and moral responsibility for children.

I start by briefly reviewing some ways in which responsibility for children has been conceptualized philosophically. I then look at the tension between biological parenthood and social recognition of parental status (and, implicitly, responsibility for children). I analyse the implications of findings in epigenetics for the ascription of biological parenthood, and explore broadly shared understandings of procreative responsibility, assessing its extension to include all (individual or collective) actors that determine a child's biology. Throughout the chapter, I problematize the focus on genetics and biology in the ascription of moral responsibility for children, using the example of epigenetics as a crossover between social and genetic factors that contribute to a child's life. By the end of the chapter, I aim to show that 'it takes a village' to make a child who that child is. 'Biological parents' (whatever that means) may be the ones who bring a child into this world. However, much of the child's life will depend, not only socially but also biologically, on the experiences and choices of many more people as well as on a host of other circumstances, many of which are beyond these people's control. Ascription of responsibility for children needs to reflect this complexity.

Moral responsibility for children

As we have already seen in this volume, there are many ways of conceptualizing and understanding moral responsibility.[1] Moral responsibility may be prospective: for example, we may say that a person becomes responsible for a child as a result of taking on a care-taking role. Responsibility may also be retrospective: one may say, for example, that someone who participated in the creation of a child is thereby responsible for that child. When we talk about responsibility, we may mean one or the other of these, or both. In relation to children specifically, the question of who has moral responsibility may be raised in order to determine either retrospective responsibility or prospective responsibility for them. These are different questions.

Both morally and legally, children's interests and vulnerability are the basis for responsibilities held by other moral agents. Throughout the Western world, the primary holders of moral responsibility for children are, by default, those who are recognized as their parents. Because parental responsibilities are often codified into laws, they are ascribed to whoever is the child's legal parent, which then extinguishes 'competing' responsibilities on the side of those who are not the legal parents. Parent–child relationships are commonly seen as binary and exclusive: it is the legal parent who bears parental responsibility, and (in most legislatures) no more than two adults can be a child's legal parents. If one is not the parent, one has at most some temporary and well circumscribed responsibilities brought on by one's role, such as that of a nanny or a teacher. However, the parents have control over these relationships: for example, they can fire the nanny or move the child to another school. This status quo rests on the assumption that parental responsibility trumps any other responsibilities for specific children. Of course, this does not mean that parents' decisions cannot be questioned. Children have rights and are entitled to protection by the state, even from their own parents. But, unless there are serious reasons to suspect that parents have acted severely against their children's interests, their discretion in a variety of matters will typically go largely unchallenged. How strong these reasons must be, and how severe the actions must be that trigger state interventions, differs significantly between countries.

One feature of the recognition and formalization of responsibilities for children is the conviction that children belong with their biological parents, and that therefore responsibility for children rests primarily with the biological parents. This conviction survives despite being contested socially, ethically and legally (Cutas and Chan, 2012). Innovations in human reproduction challenge this assumption. Practices such as gamete and embryo donation have increased the number of children who are born into families with whom they do not share a genetic link. This makes it more complex to determine who should be allowed to develop or maintain relationships with these children. Furthermore, developments in the justification of parental rights also call into question the presumption in favour of the biological parents: if parental rights are grounded in the interests of children, then genetic connections are no longer central – or are altogether irrelevant – unless it is in the interests of the children that they are so recognized. Meanwhile, in some European countries, regulations in areas such as immigration use DNA as evidence of parent–child relationships. In some US states, men can demand DNA testing and have their genetic parentage acknowledged against the wishes of the husband of the mother and the child's legal father (Carbone and Cahn, 2011; Smajdor and Cutas, 2014). So, while in some ways, Western societies are moving away from biological accounts of parenthood, in others they reinforce them.

Gamete donors and other participants in fertility treatments have often been represented as simply providing a service, product or treatment. However, it has been argued that we should subject these contributions to a broader notion of responsibility that may be procreative but not parental (Fahmy, 2013). Such an endeavour allows examination of procreative responsibility independently of parental responsibility. The people who contributed to bringing the child into existence thereby acquire responsibility for that child whether their contribution was biological or not, and whether they are to be recognized as the child's parents or not. This distinction between procreative and parental responsibility will also be useful when, later in this chapter, we look at biological influences onto children's lives that are neither parental nor procreative but may shape children's biology.

A related distinction in discourses on moral responsibility for children is that between primary and secondary responsibilities (Macleod, 2007). While parents have primary moral responsibility for their children, other parties may have secondary responsibilities for them. These parties may include not only gamete donors and other individual participants in the creation and life of the children but also collective units such as schools, hospitals or states. This allows the parents to function as core decision-makers on behalf of their children, while other individuals or groups may have concurrent, but more diluted, responsibilities for them. In practice, what happens when the exercise of secondary responsibilities for children clashes with parental responsibilities depends on the legislature. In some countries, such as the UK, going against parents' decisions, when these are deemed to be against the child's interest, is a fairly straightforward process. In others, there is much more deference to parental authority even where there are good reasons to believe that the parents are acting against a child's interests (Wilkinson and Savulescu, 2018).

Other sources of responsibility for children may ensue from relationships with children that have led them to form attachments. People's personal and social connections are perceived as much more fluid than genetic or family relationships (Braithwaite, 2010; Brake, 2012). Expectations that people behave in certain ways, or are warranted certain protections of their relationships, are much stronger in the case of parent–child relationships. Parents are entitled to exclude other people from their children's lives arbitrarily, and in general they enjoy comprehensive privileges in relation to their children, regardless of the children's or anyone else's interests (Bartlett, 1984; Gheaus, 2017). This exclusivity renders invisible some connections and attachments that may be extremely important for the children and for adults who are not their legal parents. While the exclusivity is increasingly being challenged,[2] it is still pervasive. Although it may help to simplify the exercise of societal responsibility for children, by placing it almost entirely with the legal parents, this may not be compatible with current views regarding the moral status of children and the importance of their interests.

Changes in views on the moral status of children and the conceptualization of responsibility for them have significant implications not only for parent–child relationships but also for the relationship between parents as well as that between parents and other parties, including society in general. Together with changes in patterns of relationships between adults and expectations of and from parents, these call for a restructuring of relationships between parents. Co-parenting is gradually replacing marriage as that which binds parents and generates lasting responsibilities, not only for their children, but for each other as sharers in parenting (Cook, 2012; Cutas and Hohl, 2021). This restructuring of adult relationships also leads to the question of whether parents should allow, encourage and nurture the connections that their children may have with others, and organize their own lives accordingly.

Some authors go even further than this to question the very use of adult perspectives when analyzing moral responsibility for children. For example, Wiesemann (2016) argues that what consolidates the duty to recognize children as moral agents is their trust and vulnerability, rather than their (potential for) autonomy as defined by adults. She uses the term 'moral adultism' to describe the tendency to 'translate' children's lives and interests in terms of adults' interests and adult values, or possibly children's future interests as adults (which may even be given precedence over their interests as children). Moral adultism, argues Wiesemann, stands in the way of actually, substantially, recognizing children as the moral equals of adults. Rather than seeing them and respecting them for who they are, we are projecting our own, adult, perspective onto them.

In short, responsibility for children may be prospective or retrospective, parental or non-parental, parental or procreative, maternal or paternal, primary or secondary, individual or collective. While responsibility for children tracks biological contributions to children's lives, it also responds to the recognition and promotion of the types of relationships between adults. At the same time, changing legal and moral conceptions of children's moral status and the justification of adults' claims to children also change the basis for the recognition of relationships with children and the ascription and content of responsibilities for them.

Biology, responsibility and parenthood

A woman and a man love each other very much, get married (to each other!), and together have one or several children. This is a family. The woman and the man then are the parents of the children they have created and are responsible for them. The children will display a combination of their parents' traits and genetic potential. Sure, some children arise out of a more fortunate combination of gene pools than others, but such is life. Restricting people's liberty to 'found a family' is associated with eugenics and has a particularly dark recent history. It is accepted – and indeed seen as self-evident throughout the Western world – that people should make their own decisions about who they want to reproduce with, and that they are responsible for the children resulting from these unions. There are only a few situations in which there may be an expectation that these choices take into account the impact on children. These situations include serious conditions that would dramatically affect the children's potential to flourish. Beyond such extreme cases, people are able to make their decisions freely, and it is only their post-conception decisions that may be questioned.

Once the child is conceived, the woman and the man are not always seen as equally responsible for the children they have together. The biological differences between women's and men's contribution to children's existence have been taken to imply that women have far greater moral responsibility for their future children. Against the background of progress in medicine and genetics, pregnancy has made the female body vulnerable to increasingly far-reaching demands (Smajdor, 2011; Kukla, 2016). Women are expected to refrain from anything that could possibly harm their children before they are born or even conceived. In the UK, all expectant mothers are encouraged to get tested for carbon monoxide to detect whether they smoke (Gregory, 2019). Guidelines urge women who are planning a pregnancy, as well as expectant mothers, not to consume alcohol at all (Department of Health, 2016). That women may conceive and be pregnant is also one of the reasons why women have been excluded from many types of medical research for a long time. The same standards tend not to be applied to men. The question of whether fertile men should refrain from drinking alcohol or smoking, in order to avoid risks to potential children that they may have, is so striking that it has recently made the rounds on social media as a sarcastic, humorous proposal. However, lifestyle factors including smoking or the consumption of alcohol also have an impact on male reproductive tissue, and do so early on in life. While there is research connecting the quality of male reproductive tissue with health risks to the child, this evidence tends to be absent in the policing of reproductive choices, which primarily affects women (Hens, 2017).

Legally, the story of the woman and the man, and the family they found together, can diverge from the facts of biological (and especially genetic) reproduction. The script lives a life of its own: by default, it is the woman and her husband who form a sanctioned family form. The biological father may be someone else. Or the couple may have had IVF and (1) egg donation, (2) sperm donation, or (3) both, and so possibly neither is genetically related to the child. Or they may have had their child(ren) with help from a surrogate mother, in which case it is the man and the surrogate mother who are the biological parents (if the surrogate mother also contributed the egg) or the three of them (if the wife contributed the egg), or the man and the surrogate mother and the egg donor (if someone else contributed the egg), or the surrogate mother and the egg donor and the sperm donor. If mitochondrial donation[3] is also involved, then yet another person may also have contributed biologically.

Although legal default parenthood and biological procreation need not completely overlap, biological contribution and parenthood are connected in the parenthood script that many of us operate with either consciously or not. A liberalization of parenthood status to include all these types of connections simultaneously is not in sight. This would clash with another firmly held expectation about parenthood: that every child can have no more than two parents. As we have seen in the previous section, ideas of moral responsibility for children are being adjusted in light of a broader understanding of children's moral status and the many ways in which children have been unjustifiably undervalued (see also Gheaus et al, 2019). However, this has not yet led to children no longer being attributed to pairs of adults and to relationships between adults no longer determining parental status.

The idea that biology entails parenthood and the idea of the legitimate (married) couple as the model for respectable parenthood have in common the view that child(ren) are in one way or another generated by their parents. The children are made either from others' biological contributions or from their parents' reproductive projects. Ideally, these coincide: the birth mother is the genetic mother and the biological father is her husband. The expectation that children are made by their biological progenitors and that this fact forms the basis of claims over them (such as parenthood) is subject to a number of complications, not only, as we see in the next section, from new understandings of epigenetics, but also from other biological possibilities. These include the splitting of biological motherhood into two, made possible by surrogate motherhood and embryo transfer. More intriguingly, it has been found that DNA from (male) fetuses travels to and remains in their gestational mother's brain long after pregnancy (Chan et al, 2012). Scientists speculate that this 'colonization' of the mother is not accidental and has a purpose: the benefit of the fetus (Boddy et al, 2015). If this is true, and if contributing to the biology of a person is parenthood, then the fetus may be said to become his mother's parent. Or, if we are to abstain from using parenthood language, he will become a biological contributor to his own mother.

Where does epigenetics fit in?

Epigenetics, by bringing into the foreground the relationship between the environment in which a child develops (including the uterus) and the way that their genes are expressed, risks increasing the divide between women's and men's perceived responsibilities for their children. As women are children's prenatal environment, epigenetics may be – and has been – seen as providing further ammunition to extend maternal responsibilities for children to before birth and even before conception. Richardson et al (2014) have warned that careless reporting of epigenetic influences may lead to harm to women, as they may be blamed for epigenetic effects that occur *in utero*. Likewise, Juengst et al (2014) have warned about the leap from studying pregnancy in mice to making claims about what expectant mothers should do.

These risks in the translation of epigenetics findings into moral and legal terms, with a focus on mothers' responsibilities, have been called 'epi-eugenics' (Wastell and White, 2017). Wastell and White also highlight how epigenetics may be used to expand parental, and especially maternal, responsibilities. Scientific results based on, for example, animal studies, have been translated into policies aimed at parents who are seen as under-performing with regard to their parental responsibilities: just as a stressed mother rat will neglect or hurt her pups, a stressed human mother may also hurt her children. By individualizing the causes of the distress to the pups (and babies), mothers are held responsible for conditions that may be beyond their control, while at the same time freeing policy makers themselves (and researchers who choose to subject animals to stress)– from problematizing their own contribution. Instead of unravelling the interconnectedness between structural problems and effects on parents and children, epigenetics thus 'opens new arenas for maternal responsibilisation' (Wastell and White, 2017, p 184).

By calling into question the boundaries between biology and the environment, epigenetics also confounds the distinction between biological and social parenthood. If the environment in which a child is raised influences their gene expression, in a way that is inheritable, then the environment is also a genetic contributor to the child. If contributing genetically to a child makes one a parent, then the environment in which a child is raised is also a parent of the child. Calling an environment a parent may seem counter-intuitive and perhaps incompatible with what we tend to mean when we say 'parent': especially if what we are looking for is a way of ascribing responsibility for children, we may specifically seek an identifiable moral agent, which a child's environment as such is perhaps not. Going back to the distinction between procreative and parental responsibility as a way to capture various types of responsibilities for children, one way to make sense of epigenetic input into a child's life may be to call it something like 'responsibility for shaping'. Epigenetic contributions may not be parental, and they may not be procreative, but they may be significant. Furthermore, the conditions within which children are raised are often shaped by other forces, which may themselves be determined by individual or collective agents.

We have seen in the context of mitochondrial transfer both the lure of the 'three-parent baby' discourse, and the explaining away of the donor as simply providing a little help where needed by the (only!) two prospective parents. Mitochondrial transfer involves use of the part of an egg containing mitochondrial DNA to replace the faulty mitochondrial DNA of the prospective mother. Because the nuclear DNA is that of the prospective mother, the argument goes, the mitochondrial DNA donor's contribution is not the kind that grants biological parent status (Sample, 2015). At the same time, the procedure has frequently been reported in the media as 'three-parent reproduction', and the children conceived in this way have been described as 'three-parent babies' (Hamzelou, 2016; Macrae, 2016).

However, as Anna Smajdor and I pointed out elsewhere, this suggests that genetic parenthood is a matter of degree (Cutas and Smajdor, 2018): the contribution of the egg donor in mitochondrial transfer is not enough to count as biological parenthood.

In a similar manner, one may also object to the environment or the fetus counting as a genetic contributor – or genetic parent – on the basis of degree. The argument could go like this. The environment in which a child is raised may well have an impact on their gene expression in a way that is inheritable. However, the child is essentially the same child that they would have been were they raised in another environment. They would just be weaker or stronger, taller or shorter: different predispositions would be stimulated or actualized in the various scenarios. As in the case of mitochondrial transfer, this is just a matter of degree, and does not change the essence of who the child really is (see also Chapter 4).

However, whether I have a predisposition towards obesity or am very anxious or suffer from mitochondrial disease are not marginal, negligible properties. They are very much an integral and salient part of who I am, and will have a significant impact on how I am perceived, how I navigate the world, and how much of my potential I can realize. Likewise, if it is true that fetal DNA finds its way into the gestating woman's brain, then the fetus is a genetic contributor to who she is, even if only to a very small degree. That fetal DNA may not even do anything, but it is a part of her.

Conclusion

The capacity to remove eggs from one woman's body, fertilize them and transfer them into another woman's uterus has split biological motherhood into two: genetic and gestational. Research into epigenetics indicates that the gestational mother not only helps nourish and develop the fetus in her body, but also contributes to their gene expression. In that sense, she also becomes a genetic mother: she may not transmit genes to the fetus, but she contributes to how the child's genes behave. In the same way, a rearing parent becomes a biological – and genetic – parent by also contributing to the child's gene expression. Gene expression determines a child's identity (for more on gene expression and identity, see Chapter 4).

If contributing biologically to a child's identity is parenthood, and raising a child contributes to their gene expression in significant ways that are inheritable – which means contributing biologically – then raising a child is (one kind of) biological parenthood. If it is not parenthood, it is in any case biological contribution. Whether or not I develop a life-changing disease is important to who I become. This risk depends partly on whose genetic material has made me, but may also depend on the environment that the people who made me lived in, what experiences they have had, or on what environment and what experiences I have had. Responsibility for detrimental circumstances may pertain to one's biological parents or the societies they lived in or political decisions made by others. In this sense, knowledge of epigenetics not only extends spheres of influence, but also extends the scope of collective responsibility for children's wellbeing. It thus brings closer together or altogether blurs the margins between parental, non-parental, primary, secondary, individual and collective responsibilities for children.

While epigenetics may be viewed as a basis for extending individual responsibility for children, it also reveals ways in which we are more interconnected with the world that we live in and with each other than we might like to believe. Epigenetics blurs the boundaries between biological and social parenthood. It extends but also dilutes individual moral responsibilities for children by increasing the scope of collective moral responsibility for them. In so doing, it challenges the focus on atomized individual blame and on the capacity to individually prevent or address harm to children.

Insofar as the people who raise a child determine the environment around that child, they are also biological contributors to the child. Insofar as the conditions in which a child is raised are determined by other factors, such as societal inequalities that condemn some people to living in conditions that make adequate development unlikely or difficult, then such inequalities – and the forces that cause them – are morally problematic, just as a parent's failure to safeguard their child's wellbeing is problematic. As these forces are systemic, solutions are also systemic: we cannot fix systemic problems by castigating individual parents whose choice of world in which to conceive and bring up a child is limited. In short, human reproduction and childrearing really do 'take a village'.

Notes

1   This section of the chapter is a further development of a part of Cutas (2021).
2   There have been cases, in the Western world, in which grandparents have obtained visitation rights of their grandchildren, regardless – and despite – the wishes of parents (Henderson, 2005).
3   Mitochondrial transfer involves the removal of an egg's nucleus which is then placed into another egg. The typical reason for this procedure is the presence of mitochondrial disease: by removing the outer shell of the egg, the risk of the baby being born with mitochondrial disease is removed. The baby resulting from that 'new' egg will inherit the mitochondrial DNA of the egg donor.

Acknowledgements

I am grateful to Anna Smajdor, Emma Moormann, Kristien Hens and Maria Hedlund for helpful comments. Work towards this chapter was supported by the Marcus and Amalia Wallenberg Foundation (grant number 2020-0074).

References

Bartlett, K. (1984) 'Rethinking parenthood as an exclusive status: the need for legal alternatives when the premise of the nuclear family has failed', *Virginia Law Review*, 70(5): 879–963.

Boddy, A., Fortunato, A., Wilson Sayres, M. and Aktipis, A. (2015) 'Fetal microchimerism and maternal health: a review and evolutionary analysis of cooperation and conflict beyond the womb', *BioEssays*, 37(10): 1106–18.

Brake, E. (2012) Minimizing Marriage. *Marriage, Morality, and the Law*, Oxford: Oxford University Press.

Braithwaite, D.O., Bach, B.W., Baxter, L.A., DiVerniero, R., Hammonds, J.R., Hosek, A.M., et al (2010) 'Constructing family: a typology of voluntary kin', *Journal of Social and Personal Relationships*, 27: 388–407.

Carbone, J. and Cahn, N. (2011) 'Marriage, parentage and child support', *Family Law Quarterly*, 45: 2.

Chan, W.F., Gunrot, C., Montine T.J., Sonnen, J.A. and Nelson, J.L. (2012) 'Male microchimerism in the human female brain', *PLoS One*, 7(9): e45592.

Cook, P. (2012) 'On the duties of shared parenting', *Ethics and Social Welfare*, 6(2): 168–81.

Cutas, D. (2021) 'Etica relației dintre copii, părinți și stat', in A. Volacu, D. Cutas and A. Miroiu (eds), *Alegeri Morale: Teme Actuale de Etică Aplicată* , Iași, Romania: Polirom.

Cutas, D. and Hohl, S. (2021) 'In it together? An exploration of the moral duties of co-parents', *Journal of Applied Philosophy*, 38(5): 809–23.

Cutas, D. and Smajdor, A. (2018) 'Reproductive technologies and the family in the XXIth century', in S. Giordano, J. Harris and L. Piccirillo (eds), *Bridging the Gap between Science and Society: A Second Anthology on Freedom of Scientific Research*, Manchester: Manchester University Press.

Cutas, D. and Chan, S. (eds) (2012) *Families: Beyond the Nuclear Ideal*, London: Bloomsbury Academic.

Department of Health (2016) *Alcohol Guidelines Review. Summary of the Proposed New Guidelines*. Available from: www.assets.publishing.service. gov.uk/government/uploads/system/uploads/attachment_data/file/489 795/summary.pdf [Accessed 1 August 2023].

Fahmy, M. (2013) 'On procreative responsibility in assisted and collaborative reproduction', *Ethical Theory and Moral Practice*, 16: 55–70.

Gheaus, A. (2017) 'Children's vulnerability and legitimate authority over children', *Journal of Applied Philosophy*, 35(1): 60–75.

Gheaus, A., Calder G. and de Wispelare, J. (eds) (2019) *The Routledge Handbook of the Philosophy of Childhood and Children*, London: Routledge.

Gregory, A. (2019) 'All mothers-to-be will be tested for smoking', *The Sunday Times* [online], 30 March. Available from: thetimes.co.uk/article/ all-mothers-to-be-will-be-tested-for-smoking-zcr6gvjsm [Accessed 1 August 2023].

Hamzelou, J. (2016) 'World's first baby born with new "3 parent" technique', *New Scientist* [online], 27 September. Available from: www.newscientist. com/article/210721 9-exclusive-worlds-first-baby-born-with-new-3-parent-technique [Accessed 1 August 2023].

Henderson, T. (2005) 'Grandparent visitation rights: successful acquisition of court-ordered visitation', *Journal of Family Issues*, 26(1): 107–37.

Hens, K. (2017) 'The ethics of postponed fatherhood'. *International Journal of Feminist Approaches to Bioethics*, 10(1): 103–18.

Juengst, E. Fishman, J., McGowan, M., and Settersten Jr., R. (2014) 'Serving epigenetics before its time', *Trends in Genetics*, 30(10): 427–9.

Kukla, R. (2016) 'Equipoise, uncertainty, and inductive risk in research involving pregnant women', in F Baylis and A Ballantyne (eds), *Clinical Research Involving Pregnant Women*, Dordrecht: Springer.

Macleod, C. (2007) 'Raising children: who is responsible for what?', in S. Brennan and R. Noggle (eds), *Taking Responsibility for Children*, Waterloo, Ontario: Wilfrid Laurier University Press.

Macrae, F. (2016) 'Britain's first three-parent baby could be born within one year: scientists say the controversial IVF technique is ready for use', *Daily Mail* [online], 9 June. Available from: www.dailymail.co.uk/sciencetech/arti cle-3631770/Britain-s-three-parent-babyborn-ONE-YEAR-Scientists-say-controversial-IVF-technique-ready-use.html [Accessed 1 August 2023].

Richardson, S., Daniels, C., Gillman, M., Golden, J., Kukla, R., Kuzawa, C., et al (2014) 'Society: don't blame the mothers', *Nature*, 512(7513): 131–32.

Sample, J. (2015) '"Three-parent" babies explained: what are the concerns and are they justified?', *The Guardian* [online], 2 February. Available from: www.theguardian.com/science/2015/feb/02/three-parent-babies-explained [Accessed 1 August 2023].

Smajdor, A. (2011) 'Ethical challenges in fetal surgery', *Journal of Medical Ethics*, 37(2): 88–91.

Smajdor, A. and Cutas, D. (2014) 'Artificial gametes and the ethics of unwitting parenthood', *Journal of Medical Ethics*, 40: 748–51.

Wastell, D. and White, S. (2017) *Blinded by Science: The Social Implications of Epigenetics and Neuroscience*, Bristol: Policy Press.

Wiesemann, C. (2016) *Moral Equality, Bioethics, and the Child*, Dordrecht: Springer.

Wilkinson, D. and Savulescu, J. (2018) *Ethics, Conflict and Medical Treatment for Children: From Disagreement to Dissensus*, London: Elsevier.

6

AI and Epigenetic Responsibility

Maria Hedlund

Introduction

Since epigenetics became a subject of interest for social science and the humanities about a decade ago, questions about responsibility have been a core focus. This is not surprising. While changes in the DNA are unpredictable, epigenetic changes, although complex, are in principle possible to track to their sources. This means that – at least in some cases – they may be connected to individuals who, in a causal respect, may be seen as responsible for them. Hence, the main responsibility issue that epigenetics gives rise to is expansion of the scope of responsibility attribution. Lifestyle factors accentuate individual responsibility, whereas environmental factors draw attention to collective responsibility. Norms of justice play a key role in discussions on how to distribute responsibility for epigenetic effects between individuals and collectives such as corporations and the state (Hedlund, 2012). Such issues have stimulated calls for further nuance in the responsibility debate, for instance by suggesting a need to go beyond comparisons between epigenetics and genetics or by drawing attention to the fact that how we talk about epigenetics may imply different distributions of responsibility (Dupras and Ravitsky, 2016a,b). These and other contributions are welcome to scholars addressing ethical, legal and social implications of epigenetics, and give important refinements to the responsibility discussion. However, paying attention to the main differences between epigenetics and genetics – while keeping in mind the subtleties – gives room to elucidate the principal and intricate responsibility relationships between individuals and collectives, and between current, past and future generations, that epigenetics brings to the fore.

Now, developments in the field of artificial intelligence (AI) have further complicated the issue. Machine learning (ML), a subfield of AI by which the machine learns to recognize patterns in big datasets, enables analysis of complex pathways in medical data for diagnostic, prognostic and therapeutic purposes (Holder et al, 2017; Hamamoto et al, 2019; Brasil et al, 2021). This is of crucial importance for progress in the analysis of epigenetic data (Rasuchert et al, 2020). AI technology in general, and ML in particular, give rise to very special responsibility issues. These systems learn by experience (of data patterns) and improve their prediction capacity over time, and the designers and programmers cannot always explain how the system reached its conclusion (Campolo and Crawford, 2020). While it may be argued that humans always have responsibility for what the machines that they created do, this diminution of control is sometimes referred to as a responsibility gap (Matthias, 2004; Gunkel, 2017) – a space in which no one can be ascribed responsibility. Even though the notion of a responsibility gap is contested (Köhler et al, 2017), the phenomenon that it refers to gives the question of epigenetic responsibility a further dimension. But this is not the only way in which AI technology and its deployment have implications for responsibility attribution. Machine learning is basically a statistical process, which means that data are at centre stage. Aspects such as the number of data points (Kwon, 2020) and the representativeness of datasets (Cirillo and Rementeria, 2022) are very important for the accuracy of data analysis, which is also important for the discussion on responsibility ascription. AI technology is a technically complicated field that is difficult to understand for those who are not themselves experts on the technology (Hedlund and Persson, 2022). In combination with epigenetics, another technically complicated field, this makes responsibility attribution even more intricate.

The aim of this chapter is to disentangle the increasing complexity of responsibility relationships that epigenetics gives rise to when AI is added to the picture. As with the emergence of epigenetics, AI development is seeing a multifaceted scholarly discussion on responsibility. And, as with epigenetics, AI technology gives rise to questions concerning responsibility that not only urge society to reflect upon the consequences of various responsibility attributions, but also bring further dimensions to theorizing around responsibility. When AI technology is used to analyse epigenetics, additional complexity enters the responsibility equation. In this chapter, I aim to elucidate some of this complexity and show how it contributes to the discussion on epigenetic responsibility. I first discuss some categories of responsibility, and relate these to questions in the ongoing discussion on epigenetic responsibility. Next, I briefly introduce AI and highlight some intriguing questions of responsibility attribution that AI raises. I then illustrate how some of these questions are brought to the fore in medical AI practices. Finally, I conclude that introduction of medical AI analysis increases the responsibility of various experts who need to collaborate. At the same time, AI analysis of epigenetic data is emerging as a valuable tool for personalized medicine, or precision medicine, that takes into account individual variability in genes, environment and lifestyle to tailor preventive and therapeutic strategies to individual patients (Hamamoto et al, 2019; Catura-Solarz et al, 2022). By internalizing socio-environmental determinants of health, the responsibility for health is increasingly put upon the individual (Dupras and Ravitsky, 2016b; Chiapperino and Testa, 2016). These concurrent trends require further research.

Categories of responsibility

To start with, it should be noted that I am only discussing moral responsibility here, not legal responsibility. Moreover, I only address aspects of moral responsibility that apply to the discussion of how AI affects epigenetic responsibility, namely forward-looking responsibility, the relationship between individual and collective responsibility, and, to some extent, the related problem of 'many hands', referring to situations when only individuals acting jointly can ensure a certain outcome.

While backward-looking responsibility is undoubtedly important, in this context I focus on forward-looking responsibility, dealing with the question of who should ensure that something desirable happens or that something undesirable does not happen. To be responsible in a forward-looking way, you need to have the capability to bring about some desirable state of affairs (or prevent some undesirable state of affairs) (Smiley, 2014; Van de Poel, 2015a).[1] Whereas for backward-looking responsibility, causal connection to past events is essential, for forward-looking responsibility it is more pertinent to talk about efficacy: the ability to produce a desired result (Graafland, 2003; Van de Poel, 2015a). However, efficacy does not tell us anything about which agent responsibility should be ascribed to. Distribution of responsibility between possible agents requires some normative basis, such as justice or fairness. Depending on context, this may mean different things. For instance, in the context of climate change mitigation, the contribution principle ('the polluter pays') and the benefit principle (that those who gain from something that is harmful to others should pay) are often discussed. In the context of epigenetic responsibility, the principle of position has been put forward: those who are in a position to make a difference – because they have relevant resources or because they are at the right place at the right time – have a moral obligation to do so. Position, so understood, may, but need not, overlap with capacity (Hedlund and Persson, 2022). This argument may point to a societal responsibility to mitigate adverse effects of epigenetic mechanisms (Hedlund, 2012). As has been extensively pointed out (Meloni and Müller, 2018; Valdez, 2018; Meurer, 2021), attributing epigenetic responsibility to individuals may be stigmatizing, as well as counter-productive, due to structural and other constraints on the capacity or willingness of individuals to change their behaviour. This paves the way to collective epigenetic responsibility, attributing (some) responsibility for individual health to collectives such as the state, healthcare institutions and companies. The call for a focus on collective responsibility in contexts where the scope for vulnerable individuals to act is constrained has gained a lot of support (for example, Pentecost and Meloni, 2018; Meurer, 2021; Santaló and Berdasco, 2022; see also Chapter 2). However, it has also been pointed out that the moral basis for collective epigenetic responsibility is vulnerable to the same criticism as that for individual epigenetic responsibility (Chiapperino, 2020; see also Chapter 3).

Certainly, collective responsibility is a disputed concept. A necessary requirement for moral responsibility is moral agency. Adults, for example, are typically regarded as having moral agency. Collectives are trickier in this regard. It is not completely clear to what extent a collective may qualify as a moral agent. For one thing, collectives cannot care about the outcome of their actions like individual sentient beings do (Hedlund and Persson, 2022). Moreover, collectives do not have the kind of control over their actions that is crucial for moral responsibility (Held, 1970; Hakli and Mäkelä, 2019). However, under some conditions, it is reasonable to regard collectives as moral agents. Groups that are 'organized and capable of carrying out projects in a purposeful action' (Smiley, 2014, p 4), such as states, companies and universities, meet the requirements of moral agency, meaning that they can be responsible as a single body (Thompson, 1987; Van de Poel, 2015b; Hedlund, 2022).

As noted earlier, backward-looking responsibility is important in ascribing responsibility to collectives whose actions or inactions have contributed to harmful epigenetic effects. However, from a normative perspective, it is also important to focus on what could be done to mitigate adverse epigenetic effects, to take a forward-looking perspective (Smiley, 2014). The ascription of collective responsibility is vulnerable to criticism about the coherence of collectives and thereby the possibility of collective agency, as described earlier. However, Chiapperino (2020) argues that efficacy and solidarity are in fact an effective basis for ascribing epigenetic responsibilities to collectives. It is precisely because no-one (individual or collective) can have full control that we need collective actors such as the state and its public institutions to take on (epigenetic) responsibility. In a sense, this is what we have collective actors (such as the state) for. Although there certainly are some cracks in the moral and/or conceptual foundation of collective responsibility (Chiapperino, 2020), forward-looking collective responsibility is necessary to bring about certain states of affairs that no individual could achieve on their own, when coordinated collective action is necessary to avoid future harm (Van de Poel, 2015b).

One question that the notion of collective epigenetic responsibility raises is that of the role of epigenetics, or, more specifically, the role of knowledge about epigenetic mechanisms: does collective responsibility make epigenetics irrelevant? The study of epigenetics has revealed the molecular relationship between environmental exposures and diseases, and has expanded our knowledge of what happens in gene–environment interactions at the molecular level. However, awareness of the importance of lifestyle and environment for health outcomes is far from new (Hedlund, 2012; Chiapperino and Panese, 2019). Neither is the claim that society should take responsibility for the wellbeing of its citizens; the debate about where to draw the line between individual and societal responsibility, or between the private and the public sphere, has been a core concern in political and philosophical thinking for centuries (Bexell, 2005). Moreover, the development of the welfare state, which takes on responsibility for the well-being of all citizens – especially those who are worse off (Aravacik, 2019) – has taken place independently of emerging knowledge concerning epigenetic mechanisms. The point here is that the notion of collective responsibility is not particular to epigenetic effects.

However, in light of the development of personalized medicine (Holder et al, 2017; Santaló and Berdasco, 2022), and in a biopolitical landscape that concentrates on individual clinical interventions to mitigate structural inequalities (Dupras and Ravitsky, 2016b), epigenetic responsibility seems increasingly to be imposed upon the individual. This tendency is further enabled and facilitated by the development of analytical methods based on AI technology. The possibility of using AI to analyse epigenetic data offers promising new avenues for the development of predictive tools for many diseases. One example of such a disease is adolescent idiopathic scoliosis, for which epigenetic factors seem to play a crucial role (Faldini et al, 2022).

Responsibility issues that AI gives rise to

AI technology, and more specifically ML, with its capacity to see patterns in big data, has strong potential in data-driven medical fields such as epigenetics. AI technology may help physicians in interpreting information-rich clinical data, and may even be essential for understanding epigenetic processes (Holder et al, 2017; Rauschert et al, 2020). AI is a powerful resource for deciphering complex epigenetic marks at the genome-wide level (De Riso and Cocozza, 2021). By 'learning' from data and previous behaviour, ML tools continue to improve their accuracy at investigating the epigenome as well as expanding the medical applications of epigenetic-based disease diagnosis. For instance, ML analysis of epigenomics will be useful for molecular diagnostics for specific diseases, such as cancer, neurodegenerative disorders, fertility issues or obesity (Holder et al, 2017). The use of AI technology for the analysis of epigenetic as well as genetic data may spur therapy development not only for common diseases such as cancer, but also for rare diseases (Brasil et al, 2021). ML enables multimodal analysis of big omics data, contributing to the integrated understanding of genetic variation and epigenetic deregulation that is critical for the development of precision medicine, or personalized medicine (Hamamoto et al, 2019).

Although there is no universally agreed definition of AI, it may generally be described as non-organic cognitive systems that can think and act rationally and similarly to humans (Russell and Norvig, 2010).[2] More specifically, AI may refer to 'systems that display intelligent behaviour by analyzing their environment and taking actions — with some degree of autonomy — to achieve specific goals' (European Commission: High-Level Expert Group on AI, 2019).[3] These understandings of AI underscore the ability of AI systems to 'learn' and to act autonomously, as well as the fact that current AI systems are task-specific.[4] A characteristic of ML algorithms is their capacity to see patterns in big data and to make predictions and classifications (Busuioc, 2020). As performance of any AI system depends on its data, the quality of the data used is of utmost importance (Cruz Rivera et al, 2020). For instance, it is important that the data used to 'train' the algorithms are unbiased so that bias in the predictions is avoided (Akter et al, 2022), and that the data are correctly labelled (for example, disease/ not disease) (Rauschert et al, 2020).[5] Of relevance for our discussion on responsibility is also the learning capacity. As machines learn, algorithms adapt and drift away from the original model (Russell and Norvig, 2021). This is partly what gives rise to the complexity and opaqueness of ML algorithms and what gives them their 'black-box' character (Rudin, 2019). AI systems do not have a theoretical understanding of underlying causal relationships, but make their predictions by brute pattern recognition in big datasets. This may be particularly challenging with analyses of epigenetic data. In their recent work, Santaló and Berdasco (2022) have referred to causality as the Achilles heel of epigenetic research. In many cases, it is not possible to determine whether an epigenetic disturbance is a cause or a consequence of disease (Berdasco and Esteller, 2019). Later, I explore how the use of AI may complicate epigenetic responsibility, but let us first have a look into three well-recognized issues with AI that give rise to difficulties with ascription of responsibility: machine autonomy, bias and discrimination, and algorithms as black boxes.

Autonomy is a feature of AI technology with implications for responsibility attribution. In this context, autonomy refers to machines that make decisions and act without the direct involvement of humans. However, this does not mean that humans have not been involved and decided what the machine should do, only that execution of the task takes place without human guidance (Jain and Prathiar, 2010). While it is uncontroversial to claim that the machine lacks moral agency and thus responsibility should rest with some human(s) (Hedlund, 2020), the question remains as to which human(s) should have that responsibility: the developer, the programmer, the retailer, the user?

When we discuss responsibility for AI at the individual level, it is necessary to consider the kinds of individuals that may be brought into question. These may include experts, policy makers, providers or users. On the basis of their expert knowledge about AI technology, designers of AI systems and ML algorithms have a responsibility to make sure the systems and algorithms work as they are supposed to do, and they must not be negligent about the technical details. This is not contentious, and comes with their professional responsibility (Muyskens, 1982; Davis, 2012). Less self-evident is whether these agents also have a responsibility for the consequences of the AI systems and ML algorithms when they are put into use, and if so, whether this responsibility covers the outcomes and the consequences of these outcomes of the algorithms they develop (Hedlund and Persson, 2022). Policy makers obviously have a responsibility for the legal frameworks of AI development and deployment. For example, they need to protect against the risk that fundamental rights may be breached due to biased outputs generated by AI systems (European Commission, 2021). While possibly less relevant for medical and epigenetic analysis, ordinary users may also have responsibility for AI-governed systems. For instance, as individual search patterns affect how search engines present their results (Haider and Sundin, 2020), individuals themselves have some (causal) responsibility for the search results that turn up in their search engines. Whether they are also morally responsible is less clear (Persson and Hedlund, 2021).

As noted earlier, another aspect of AI that has great importance for responsibility attribution is the quality of the data. Low quality may refer to different things, such as poorly labelled or inaccurate data, incomplete data, or data that reflect underlying human prejudices (Cirillo and Rementeria, 2022; Crawford, 2021). As the ML algorithms learn from the data they are trained on, features of the data such as bias concerning sex, gender, race, age, ethnicity, or other categories, will affect the predictions or decisions. For instance, a recruitment system that is trained on previous hiring decisions in a company that mainly hired men 'learns' that being a man is a successful feature and will consequently filter out applications from women (Dastin, 2018). Another example is when an AI system for digital pathology has been trained only on images of white-skinned individuals – the algorithm would perform poorly at identifying cancer on dark-skinned individuals (Buslon et al, 2022). Hence, making sure that the training data are representative is key to avoid biased decisions (Cirillo and Rementeria, 2022). Moreover, bias in AI may reflect human cognitive biases. For instance, based on the fact that the majority of AI developers are men, there is a risk that predominantly male perspectives will be built into AI design (Hedlund, 2022). Bias may thus emerge out of 'seemingly innocuous code' (Johnson, 2021, p 9945).

The problem of bias and the risk of discrimination are widely recognized and addressed in research (Agarwal and Mishra, 2021) as well as in the clinic (Tasci et al, 2022) and in policy proposals (European Commission, 2021). For questions of responsibility, it is important to recognize that biases may be generated at various stages of the lifecycle of an AI system, such as data collection and processing, model building, training and evaluation, and at the points of deployment and application in real-world settings (Cirillo and Rementeria, 2022). This means that several actors, not only data scientists, ML engineers, cognitive scientists and other AI experts, but also actors such as product owners, physicians, business analysts and policy makers, have responsibility to address the risk of bias (Agarwal and Mishra, 2021; Lekadir et al, 2021).

The ability of AI systems to learn and improve without the direct involvement of humans has importance for the discussion on responsibility, not only in relation to the data they are trained on, but also the algorithm itself. The 'learning' property of ML algorithms means that they adapt over time. To some extent, the problem with learning is the same as with autonomy (that the machine 'autonomously' executes something that a human has asked it to do), but with ML algorithms, the system may develop in a way that the human cannot foresee (Rudin 2019). This relates to the question discussed earlier, of algorithms as black boxes (Campolo and Crawford, 2020). The notion of algorithms as black boxes has led some scholars to posit the existence of a responsibility gap, suggesting that the diminution of human control of systems that learn by interaction with the environment and other agents brings about a space in which no-one is responsible (Mathias, 2004). However, the idea that learning machines give rise to a responsibility gap is disputed. Some argue that the concept of responsibility is undetermined in these discussions (Santoni de Sio and Mecacci, 2021), or that what may appear as a space devoid of responsibility is more accurately recognized as a space where responsibility is blurred (Köhler et al, 2017). Moreover, the black-box metaphor has been criticized as a way to keep business secrets and evade responsibility (Bucher, 2018; Campolo and Crawford, 2020). In addition, even if algorithmic opacity could be resolved technically, these systems could still be used irresponsibly (Hagendorff, 2021). Nevertheless, the conception of algorithms as black boxes gains a lot of attention, and we are witnessing a growing literature on how to explain the inner workings of black-box algorithms (Fjeld et al, 2020; Hagendorff, 2021). Without going into detail about the prospects of explicability (Rudin 2019), it may be noted that there are great expectations, both within the political sphere and within the AI community, that it may be possible to make algorithmic decisions understandable for users (European Commission, 2021; Gunning et al, 2021). Developing explainable algorithms is to a large extent a technical endeavour, which places much responsibility upon designers, programmers and other AI experts.

Even if it could be argued that experts on the technological aspects of AI should have a responsibility for the AI systems they develop (see Douglas, 2009), it is also important that those who make use of these systems, for instance in epigenetic analysis, take on responsibility. One example of how AI and epigenetics may jointly give rise to challenges for epigenetic responsibility is the potential use of epigenetic markers to predict adolescent idiopathic scoliosis. Adolescent idiopathic scoliosis is a progressive musculoskeletal disease that may result in cosmetic deformity, back pain, functional deficits, psychological problems and impaired social interactions. It results from the interaction of multiple genes with each other and the environment. Epigenetic biomarkers may facilitate early detection and improve risk assessment (Faldini et al, 2022). Genetic factors alone have been shown to have low predictive capacity and to be insufficient to guide clinical choices. Recent evidence (for example, Mao et al, 2018; Shi et al, 2018) suggests epigenetics as a more promising field for the identification of factors associated with progression of adolescent idiopathic scoliosis. By making use of data not only from spine radiographs and clinical observations, but also genetic and epigenetic factors such as methylation status (Meng et al, 2018; Carry et al, 2021), AI may be used to develop a predictive tool for adolescent idiopathic scoliosis, which may facilitate the development of therapeutic strategies based on genetic and epigenetic factors (Faldini et al, 2022). This example illustrates a case where questions of epigenetic responsibility may be interwoven with responsibility issues connected with AI. Environment and thereby epigenetic factors do play a role, which brings up the question of the extent to which the affected individual is responsible for their own condition. At the same time, AI technology is used to analyse these factors, raising the question of how to distribute responsibility between actors such as AI developers and the medical users of the AI system. Moreover, as will be demonstrated, the relationship between the doctor and the patient could be affected. In the following, I outline some situations in which application of AI in a medical setting generates responsibility issues that may complicate notions of epigenetic responsibility.

How does AI affect epigenetic responsibility?

Although clinical epigenetics is still in its infancy (Berdasco and Esteller, 2019), the prospects of using AI for epigenetic analysis are great (Rauchert et al, 2020). However, given the responsibility issues that AI technology gives rise to, it may be expected that the increasing use of AI technology for epigenetic analysis complicates responsibility ascription. This would apply across several domains, such as the collection and labelling of data, measures of data protection related to privacy and confidentiality derived from epigenetic studies, how epigenetic information affects patient autonomy (Santaló and Berdasco, 2022), the role of operators of AI systems in the clinic (Sand et al, 2021), the interaction between the physicians who use these AI systems in the clinic, the machines and the patients (Braun et al, 2020), and the debate about personal versus public responsibility for health. Moreover, contextual differences relating not only to clinical information, but also to personal, social and cultural values, affect how individuals and societies respond to medical decisions generated by medical AI (Wagner et al, 2022). As pointed out by Wagner and colleagues, the many potential areas of application of medical AI 'are paralleled by an unprecedented increase in (moral) responsibility regarding medical AI's performance and healthcare professionals' duty of care' (Wagner et al, 2022, p 111). I now look more closely at one particular aspect that has repercussions for responsibility, namely how healthcare professionals relate to AI-assisted analysis.

AI has been particularly successful in image-based diagnosis, a time-consuming endeavour in which accuracy varies between physicians, institutions and countries, and for which human error can have serious consequences (Braun et al, 2020). The high diagnostic accuracy by AI tools in the analysis of medical images has led to high expectations that the assistance of ML algorithms may increase the reliability and the speed of the diagnostic process (Grote and Berens, 2021). While much attention has been directed at technical solutions to issues with AI, implementing AI in the clinic involves concerns well beyond algorithm performance, including unintended consequences, disrupted workflows and doctor-to-patient interaction (Yu et al, 2018). Humans operating these systems are a major factor in how they are applied, and, with the introduction of AI-enabled diagnosis, the responsibilities of professions such as radiology and pathology may change considerably. Sand et al (2021) suggest that, to be competent operators of medical AI systems, physicians need to take on certain forward-looking responsibilities to meet the specific challenges that arise through medical AI.

The black-box character of AI algorithms introduces uncertainty in medical diagnosis. For instance, epigenetic markers involved in carcinogenic pathways may potentially be used for early detection of cancer by the application of AI technologies (Hamamoto et al, 2019). However, the physician may not know how the algorithm arrives at its decisions. They may choose to suggest treatments based on the AI diagnosis, and to do so, they need to trust the machine. Trustworthiness of AI systems has become an important aspect in discussions about AI development (Nature, 2018; European Commission: High-Level Expert Group on AI , 2019). For patients, trustworthiness may be about expectations of better care, and for the clinician, AI should be user-friendly and based on adequate risk–benefit analysis (Braun et al, 2020). It is also necessary that the system is robust and provides accurate output when applied in the clinic (Lekadir et al, 2021). If clinicians choose to base their recommendations on the output of the AI system, for example, in detection of epigenetic markers involved in carcinogenic pathways, they have a forward-looking responsibility to the patient to justify this choice and their reliance on the AI (Sand et al, 2021). This responsibility obviously requires that physicians have the necessary knowledge of epigenetic mechanisms and how the AI system works, and why they deem the output trustworthy, but it also requires that they are able to communicate this to the patient. While clinicians may have difficulty in explaining how a particular algorithmic output came about, their professional training and experience maybe mean that they are better placed than the patient to assess the output of opaque AI systems. In that regard, the use of AI increases the decision authority of the clinician and diminishes the ability of the patient to exercise well-informed agency (Braun et al, 2020), meaning that the ability of individual patients to take on responsibility for their own health is reduced. In line with the normative underpinnings of personalized medicine (Hamamoto et al, 2019) or the ideal and principles of shared decision-making (Braun et al, 2020),[6] this shift of responsibility towards the physician may be seen as a drawback. However, if we accept the physician as a representative of a health institution (a societal collective), we may, in accordance with the reasoning about epigenetic responsibility above, see this as a shift from (some) individual responsibility to (increased) collective responsibility.

The importance of good quality data for AI analysis is another factor that prompts the physician to take on certain forward-looking responsibilities (Sand et al, 2021). Although epigenetic data have traits that make their study amenable to ML, this does not come without challenges (Rauschert et al, 2020). One such challenge is that very large datasets are required to train the AI, especially if the diseases that are to be predicted are rare (Brasil et al, 2021). Another challenge is prediction bias. Many predictive models are biased towards populations of European ancestry, and, to avoid prediction bias, it is necessary to obtain representative datasets that do not aggravate existing health inequalities for disadvantaged populations (Rauschert et al, 2020). To be able to critically assess whether output of AI analysis are reasonable, physicians need to know what type of data is used for analysis and to have an understanding of the range of plausible outputs (Sand et al, 2021). This responsibility also requires an understanding of the limitations of the data. This is important if there is a mismatch between training data and clinical data (Challen et al, 2019), or if epigenetic datasets are not large enough for the ML algorithms to function effectively (Rauchert et al, 2020).

Although ML analysis may outperform humans in accuracy, it does so for the specific disease that it is trained for (Cruz Rivera et al, 2020). The fact that the algorithm does not detect any breast cancer cells, for example, does not preclude the possibility that the patient may have other malignancies. For the physician, it is important to be aware of this task specificity when applying the system and communicating the result to the patient (Sand et al, 2021). Again, the physician's responsibility is not only to have relevant knowledge of how the system works, but also to communicate this. When the analysis concerns epigenetic information, additional challenges are raised. As indicated by Santaló and Berdasco (2022), translating epigenetic information to a non-specialized audience requires certain communication strategies. For instance, which epigenetic information should be communicated to patients? The Bioethics Group of the International Human Epigenome Consortium suggests that, before the actual communication, the clinician should check aspects such as the accuracy of the data, the stability of the epigenetic biomarker, the causality of the epigenetic mark, and the clinical value of the biomarker (Santaló and Berdasco, 2022). However, communication is relational, and also requires human capacities such as empathy, compassion and intuition (Wagner et al, 2022). Moreover, as Santaló and Berdasco (2022) point out, the act of communicating a result based on the analysis of epigenetic data may imply a risk of imposing a burden on people with low health literacy. Health literacy often coincides with education level, socio-economic status and lifestyle, and people with low health literacy may not only have a difficulty with understanding the epigenetic result, but also reduced possibilities to change their behaviour. To this, one may add the 'burdening' of the clinician in terms of increased responsibility: not only does the clinician need to understand how the AI system works, but also how to handle risks connected with communication of epigenetic results. Santaló and Berdasco (2022) suggest that this risk may be balanced by information from epigenetics experts.

Conclusion

Although still in its infancy, we can probably expect to see AI-assisted epigenetic analysis in the clinic in the near future. While I have only hinted at some of the many aspects of AI that may impact forward-looking responsibility relationships, and potentially epigenetic responsibility, it should be clear that the distinction between individual and collective responsibility needs some fine tuning. The risks (and possibilities) of individual epigenetic responsibility are extensively elaborated in the literature, and therefore, in this chapter, the main focus has been on actors that to some extent constitute different societal collectives: clinicians and employers in healthcare institutions, and, to some extent, AI experts at universities and in corporations. Obviously, collective epigenetic responsibility does not rest on 'society' in a general understanding, but is something that certain individuals within certain collectives need to take on, and they need to do so collectively.

This chapter has illustrated one aspect of the use of AI technology for epigenetic analysis, namely how clinicians will have to take on certain forward-looking responsibilities to handle their relationship to the technology as well as to the patient. As should be clear from this illustration, it is necessary that the clinician retains the skills necessary to analyse image scans to be able to make correct judgements regarding the machine-generated results. In addition, it is necessary that clinicians acquire new skills to be able to interpret and communicate their decisions to the patient. While the responsibility to make this happen ultimately rests on their employers, the learning process will have to include experts on AI and on communication, as well as epigenetics experts.

Thus, experts of various kinds have to be involved when AI analysis is applied to epigenetics in the clinic, and they need to make their contributions in collaboration. As individuals, they contribute their part to the desired outcome, and, although each of them is individually responsible for their part, none of them is responsible for the whole, which could only be achieved collectively. This is collective responsibility in action.

As for the desired outcome, this has to be decided in inclusive forums, in which a plurality of perspectives is considered. Given the academic and political discussions on epigenetics and on AI in the last decade or so, we may extrapolate what a desired outcome could look like: good health for all, brought about by safe means, with respect for privacy and autonomy and without discrimination, especially against vulnerable groups.

Finally, it has been repeatedly emphasized that application of AI technology on epigenetic data is a prerequisite for the development of personalized medicine, or precision medicine (Hamamoto et al, 2019; Ivanovic et al, 2022). Precision medicine is part of an individualization of health, which increasingly places responsibility on the individual. Although only hinted at in this chapter, the concurrent tendencies of increased individual responsibility and increased collective responsibility that introduction of AI into medicine and epigenetic analysis gives rise to, need to be addressed in future research.

Notes

[1] This does not necessarily mean that the responsible actor themselves brings about the desirable outcome, only that they are responsible for this outcome occurring (Hedlund and Persson, 2022).

[2] While 'think' should not be equated with what we as humans do when we think, what these non-organic systems do when they process data is often metaphorically referred to as an act of thinking. In the recent edition of their textbook *Artificial Intelligence: A Modern Approach*, Russell and Norvig (2021) reject the comparison with human thinking, and describe AI in terms of rationality without specifying how rationality is enacted.

[3] It may be questioned whether 'intelligent' is an adequate description of non-organic systems, but the point is that these systems do things that appear to be intelligent (whatever the exact meaning of this term). The meaning of intelligence is a contested issue that is beyond the purpose of this chapter.

[4]   While prospects of artificial general intelligence are discussed in the AI literature (Bostrom 2014; Sotala and Yampolskiy 2015), current AI systems are specialised on one particular task.
[5]   An underlying issue here appears to be the unavoidable normativity of those data –no neutral demarcation is possible between unbiased and biased data, or between disease and not disease. So, in a way, 'quality' of the data understood as their 'objectivity' or a perfect mapping of reality can never really be achieved. This unavoidable normativity is undoubtedly worthy of its own discussion, and there is an emerging literature on this problem (see, for example, Chowdhury and Oredo, 2022).
[6]   Although there is no consensus on the exact meaning of shared decision-making in the clinic (Makoul and Clayman, 2006), a general understanding is that healthcare professionals share the best available evidence with the patients, and that patients are supported to consider the options and to arrive at an informed decision (Braun et al, 2020).

Acknowledgements

I would like to thank Daniela Cutas, Emma Moormann and Anna Smajdor for helpful comments. Work towards this chapter was supported by the Marianne & Marcus Wallenberg Foundation (grant number MMW 2018-0020).

References

Agarwal, S. and Mishra. S. (2021) *Responsible AI: Implementing Ethical and Unbiased Algorithms*, Berlin: Springer.

Akter, S., Dwivedi, Y.K., Sajib, S., Biswas, K., Bandra, R.J. and Michael, K. (2022) 'Algorithmic bias in machine learning-based marketing models', *Journal of Business Research*, 14: 201–16.

Aravacik, E.D. (2019) 'Social policy and the welfare state', in B. Açikgöz (ed), *Public Economics and Finance*, London: Intech Open, pp 3–23.

Berdasco, M. and Esteller, M. (2019) 'Clinical epigenetics: seizing opportunities for translation', *Nature Reviews Genetics*, 20(2): 109–27.

Bexell, M. (2005) *Exploring Responsibility: Public and Private in Human Rights Protection*, Lund: Department of Political Science, Lund University.

Bostrom, N. (2014) *Superintelligence: Paths, Dangers, Strategies*, Oxford: Oxford University Press.

Brasil, S., Neves, C.J. , Rijoff, T., Falcao, M., Valadao, G., Videira, P.A., et al (2021) 'Artificial intelligence in epigenetic studies: shedding light on rare diseases', *Frontiers in Molecular Biosciences*, 8: 648012.

Braun, M., Hummel, P., Beck, S., and Dabrock, P. (2020) 'Primer on an ethics of AI-based decision support systems in the clinic', *Journal of Medical Ethics*, 47: e3.

Bucher, T. (2018) *If … Then: Algorithmic Power and Politics*, Oxford: Oxford University Press.

Buslon, N., Racionero-Plaza, S. and Cortés, A. (2022) 'Sex and gender inequality in precision medicine: socioeconomic determinants of health', in D. Cirillo, S. Catuara-Solarz and E. Guney (eds), *Sex and Gender in Technology and Artificial Intelligence: Biomedicine and Healthcare Applications*, Cambridge, MA: Academic Press, pp 35–54.

Busuioc, M. (2020) 'Accountable artificial intelligence: holding algorithms to account', *Public Administration Review*, 81(5): 825–36.

Campolo, A. and Crawford, K. (2020) 'Enchanted determinism: power without responsibility in artificial intelligence', *Engaging Science, Technology, and Society*, 6: 1–19.

Carry, P.M., Terhune, E.A., Trahan, G.D., Vanderlinden, L.A., Wethey, C.I., Ebrahimi, P., et al (2021) 'Severity of idiopathic scoliosis is associated with differential methylation: an epigenome-wide association study of monozygotic twins with idiopathic scoliosis', *Genes (Basel)*, 12(8): 1191.

Challen, R., Denny, J., Pitt, M., Gompels, L., Edwards, T. and Tsaneva-Atanasova, K. (2019) 'Artificial intelligence, bias and clinical safety', *BMJ Quality and Safety*, 28(3): 231–7.

Chiapperino, L. (2020) 'Luck and the responsibilities to protect one's epigenome', *Journal of Responsible Innovation*, 7(suppl 2): S86–106.

Chiapperino, L. and Panese, F. (2019) 'On the traces of the biosocial: historicizing "plasticity" in contemporary epigenetics', *History of Science*, 59(1): 3–44.

Chiapperino, L. and Testa, G. (2016) 'The epigenomic self in personalized medicine: between responsibility and empowerment', *Sociological Review Monograph*, 64(1): 203–20.

Chowdhury, T. and Oredo, J. (2022) 'AI ethical bias: normative and information systems development conceptual framework', *Journal of Decision Systems*. https://doi.org/10.1080/12460125.2022.2062849.

Cirillo, D. and Rementeria, M.J. (2022) 'Bias and fairness in machine learning and artificial intelligence', in D. Cirillo, S. Cautuara-Solarz and E. Guney (eds), *Sex and Gender in Technology and Artificial Intelligence: Biomedicine and Healthcare Applications*, Cambridge, MA: Academic Press, pp 57–75.

Crawford, K. (2021) *Atlas of AI: Power, Politics, and the Planetary Costs of Artificial Intelligence*, New Haven, CT: Yale University Press.

Cruz Rivera, S., Liu, X., Chan, A.-W., Denniston, A.K., Calvert, M.J. and the SPIRIT-AI and CONSORT-AI Working Group (2020) 'Guidelines for critical trial protocols for interventions involving artificial intelligence: the SPIRIT-AI extension', *The Lancet Digital Health*, 2(10): e549–60.

Dastin, J. (2018) 'Amazon scraps secret AI recruiting tool that showed bias against women', Reuters [online], 11 October. Available from: https://www.reuters.com/article/us-amazon-com-jobs-automation-insight-idUSKCN1MK08G [Accessed 22 June 2023].

Davis, M. (2012) '"Ain't no one here but us social forces": constructing the professional responsibility of engineers', *Science & Engineering Ethics*, 18(1): 13–34.

De Riso, G. and Cocozza, S. (2021) 'Artificial intelligence for epigenetics: towards personalized medicine', *Current Medical Chemistry*, 30(40): 6654–74.

Douglas, H. (2009) *Science, Policy, and the Value-Free Ideal*, Pittsburgh, PA: University of Pittsburgh Press.

Dupras, C. and Ravitsky, V. (2016a) 'The ambiguous nature of epigenetic responsibility', *Journal of Medical Ethics*, 42(8): 534–41.

Dupras, C. and Ravitsky, V. (2016b) 'Epigenetics in the neoliberal "regime of truth": a biopolitical perspective on knowledge translation', *The Hastings Center Report*, 46(1): 26–35.

European Commission (2021) *Proposal for a Regulation of Artificial Intelligence: A European Approach*, Brussels: European Commission.

European Commission: High-Level Expert Group on AI (2019) *Ethics Guidelines for Trustworthy AI*, Brussels: European Commission.

Faldini, C., Manzetti, M., Neri, S., Barile, F., Viroli, G., Geraci, G., et al (2022) 'Epigenetic and genetic factors related to curve progression in adolescent idiopathic scoliosis: a systematic scoping review of the current literature', *International Journal of Molecular Sciences*, 23(11): 5914. DOI: 10.3390/ijms23115914.

Fjeld, J., Achten, N., Hilligoss, H., Nagy, A.C. and Srikumar, M. (2020) *Principled Artificial Intelligence: Mapping Consensus in Ethical and Rights-Based Approaches to Principles for AI*, Cambridge, MA: Birkman Klein Center.

Graafland, J.J. (2003) 'Distribution of responsibility, ability and competition', *Journal of Business Ethics*, 45(1/2): 133–47.

Grote, T. and Berens, P. (2021) 'How competitors become collaborators: bridging the gaps between machine learning algorithms and clinicians', *Bioethics*, 36(2): 134–42.

Gunkel, D.J. (2017) 'Mind the gap: responsible robots and the problem of responsibility', *Ethics and Information Technology*, 22(4): 307–20.

Gunning, D., Vorm, E., Yunyan Wang, J. and Turek, M. (2021) 'DARPA's explainable AI (XAI) program: a retrospective', *Applied AI Letters*, 2(4): 1–11.

Hagendorff, T. (2021) 'Blind spots in AI ethics', *AI and Ethics*, 2: 851–67. https://doi.org/10.1007/s43681-021-00122-8.

Haider, J. and Sundin, O. (2020) *Invisible Search and Online Search Engines: The Ubiquity of Search in Everyday Life*, London: Routledge.

Hakli, R. and Mäkelä, P. (2019) 'Moral responsibility of robots and hybrid agents', *The Monist*, 102: 259–75.

Hamamoto, R., Komatsu, M., Takasawa, K., Asada, K. and Kaneko, S. (2019) 'Epigenetics analysis and integrated analysis of multiomics data, including epigenetic data, using artificial intelligence in the era of precision medicine', *Biomolecules*, 10(1): 62.

Hedlund, M. (2012) 'Epigenetic responsibility', *Medicine Studies*, 3(2): 171–183. https://doi.org/10.1007/s12376-011-0072-6O.

Hedlund, M. (2020) 'När maskiner fattar beslut, vem är ansvarig?' [Decision-making machines – who is responsible?], *Statsvetenskaplig Tidskrift*, 122(4): 545–65 [in Swedish; abstract in English].

Hedlund, M. (2022) 'Distribution of forward-looking responsibility in the EU process on AI regulation', *Frontiers in Human Dynamics*, 4(Apr): 703510. https://doi.org/10.3389/fhumd.2022.703510.

Hedlund, M. and Persson, E. (2022) 'Expert responsibility in AI development', *AI & Society*. https://doi.org/10.1007/s00146-022-01498-9.

Held, V. (1970) 'Can a random collection of individuals be morally responsible?', in L. May and S. Hoffman (eds), *Collective Responsibility: Five Decades of Debate in Theoretical and Applied Ethics*, Lanham, MD: Rowman & Littlefield Publishers.

Holder, L.B., Haque, M.M. and Skinner, M.K. (2017) 'Machine learning for epigenetics and future medical applications', *Epigenetics*, 12(7): 505–14.

Ivanovic, M., Autexier, S. and Kokkonidis, M. (2022) 'AI approaches in processing and using data in personalized medicine', *arXiv*. http://doi.org/10.48550/arXiv.2208.04698.

Jain, L.C. and Pratihar, D.K. (2010) *Intelligent Autonomous Systems*, Berlin: Springer.

Johnson, G.M. (2021) 'Algorithmic bias: on the implicit biases of social technology', *Synthese*, 198(10): 9941–61.

Köhler, S., Roughley, N. and Sauer, H. (2017) 'Technologically blurred responsibility? Technology, responsibility gaps and the robustness of our everyday conceptual schema', in C. Ulbert, P. Finkenbusch, E. Sondermann and T. Debiel (eds), *Moral Agency and the Politics of Responsibility*, London: Routledge, pp 51–67.

Kwon, D.Y. (2020) 'Personalized diet oriented by artificial intelligence and ethnic foods', *Journal of Ethnic Foods*, 7(10): 10. https://doi.org/10.1186/s42779-019-0040-4.

Lekadir, K., Osuala, R., Gallin, C., Lazrak, N., Kushibar, K., Tsakou, G., et al (2021) 'FUTURE-AI: guiding principles and consensus recommendations for trustworthy artificial intelligence in medical imaging', *arXiv*. https://doi.org/10.48550/arXiv.2109.09658.

Makoul, G. and Clayman, M.L. (2006) 'An integrative model of shared decision making in medical encounters', *Patient Education and Counseling*, 60: 301–12.

Mao, S.-h., Qian, B.-p., Shi, B., Zhu, Z.-z. and Qiu, Y. (2018) 'Quantitative evaluation of the relationship between COMP promoter methylation and the susceptibility and curve progression of adolescent idiopathic scoliosis', *European Spine Journal*, 27(2): 272–7.

Matthias, A. (2004) 'The responsibility gap: ascribing responsibility for the actions of learning automata', *Ethics and Information Technology*, 6: 175–83.

Meloni, M. and Müller, R. (2018) 'Transgenerational epigenetic inheritance and social responsibility: perspectives from the social sciences', *Environmental Epigenetics*, 4(2): dvy019.

Meng, Y., Lin, T., Liang, S., Gao, R., Jiang, H., Shao, W., et al (2018) 'Value of DNA methylation in predicting curve progression in patients with adolescent idiopathic scoliosis', *eBioMedicine*, 36: 489–96.

Meurer, A. (2021) 'The end of the "bad seed" era? Epigenetics' contribution to violence prevention initiatives in public health', *The New Bioethics*, 27(4): 159–75.

Muyskens, J.L. (1982) 'Collective responsibility of the nursing profession' in L. May and S. Hoffman (eds), *Collective Responsibility: Five Decades of Debate in Theoretical and Applied Ethics*, Lanham, MD: Rowman & Littlefield Publishers, pp 167–178.

Nature (2018) 'Towards trustable machine learning', *Nature Biomedical Engineering*, 2(10): 709–10.

Pentecost, M. and Meloni, M. (2018) 'The epigenetic imperative: responsibility for early intervention at the time of biological plasticity', *The American Journal of Bioethics*, 18(11): 60–2.

Persson, E. and Hedlund, M. (2021) 'The future of AI in our hands? To what extent are individuals morally responsible for guiding the development of AI in a desirable direction?, *AI Ethics*, 2: 683–95. https://doi.org/10.1007/s43681-021-00125-5.

Rauschert, S., Raubenheimer, K., Melton, P.E. and Huang, R.C. (2020) 'Machine learning and clinical epigenetics: a review of challenges for diagnosis and classification', *Clinical Epigenetics*, 12(1): 51. https://doi.org/10.1186/s13148-020-00842-4.

Rudin, C. (2019) 'Stop explaining black box machine learning models for high stakes decisions and use interpretable models instead', *Nature Machine Intelligence*, 1(5): 206–215.

Russell, S.J. and Norvig, P. (2010) *Artificial Intelligence: A Modern Approach* (3rd edn), London: Pearson Education.

Russell, S.J. and Norvig, P. (2021) *Artificial Intelligence: A Modern Approach* (4th edn), London: Pearson Education.

Sand, M., Durán, J.M. and Jongsma, K.R. (2021) 'Responsibility beyond design: physicians' requirement for ethical medical AI', *Bioethics*, 36(2): 162–9. https://doi.org/10.1111/bioe.12887.

Santaló, J. and Berdasco, M. (2022) 'Ethical implications of epigenetics in the era of personalized medicine', *Clinical Epigenetics*, 14(1): 44.

Santoni de Sio, F. and Mecacci, G. (2021) 'Four responsibility gaps with artificial intelligence: why they matter and how to address them', *Philosophy & Technology*, 34(4): 1057–84.

Shi, B., Xu, L., Mao, S., Xu, L., Liu, Z., Sun, X., et al (2018) 'Abnormal *PITX1* gene methylation in adolescent idiopathic scoliosis: a pilot study', *BMC Musculoskeletal Disorders*, 19(1): 138.

Smiley, M. (2014) 'Future-looking collective responsibility: a preliminary analysis', *Midwest Studies in Philosophy*, 38: 1–11.

Sotala, K. and Yampolskiy, R.V. (2015) 'Responses to catastrophic risk: a survey', *Physica Scripta*, 90(1): 018001.

Tasci, E., Zhuge, Y., Camphausen, K. and Krauze, A.V. (2022) 'Bias and class imbalance in oncology data: towards inclusive and transferrable AI in large scale oncology data sets', *Cancers*, 14(12): 2897.

Thompson, D.F. (1987) *Political Ethics and Public Office*, Cambridge, MA: Harvard University Press.

Valdez, N. (2018) 'The redistribution of reproductive responsibility: on the epigenetics of 'environment' in prenatal interventions', *Medical Anthropology Quarterly*, 32(3): 425–42.

Van de Poel, I. (2015a) 'Moral responsibility', in I. Van de Poel, L. Royakkers and S.D. Zwart (eds), *Moral Responsibility and the Problem of Many Hands*, London: Routledge, pp 13–49.

Van de Poel, I. (2015b) 'The problem of many hands', in I. Van de Poel, L. Royakkers and S.D. Zwart (eds), *Moral Responsibility and the Problem of Many Hands*, London: Routledge, pp 50–92.

Wagner, N.-F., Banerjee, M. and Paul, N.W. (2022) 'Who's next? Shifting balances between medical AI, physicians and patients in shaping the future of medicine', *Bioethics*, 36(2): 111–2.

Yu, K.-H., Beam, A.L. and Kohane, IS. (2018) 'Artificial intelligence in health care', *Nature Biomedical Engineering*, 2: 719–31. https://doi.org/10.1038/s41551-018-0305-z.

7

Responsibility and the Microbiome

Kristien Hens and Eman Ahmed

Introduction

The gut microbiome is a diverse ecosystem encompassing trillions of micro-organisms, including bacteria, viruses, fungi, archaea and protozoa. It establishes a symbiotic relationship with the human host via microbiome–host interactions that occur at various levels of complexity and that are essential for maintaining bodily homeostasis (Wu and Wang, 2019). The microbiome's 'cross-talk' with the host physiology is demonstrated via its contribution to a wide range of functions, including digestion, production of metabolites, and development of the immune system (Cryan and Dinan, 2012). The gut microbiome as an ecosystem has thus emerged as a key factor in understanding human health and disease. Although the exact number is unknown, the number of microbial genes present in the human body may equal or exceed the number found in the human genome. Indeed, the genome of these microbes is sometimes called 'our second genome'(Meisel and Grice, 2017).

At this point, the reader may wonder what a chapter on the human microbiome is doing in a volume on epigenetics. There are a number of reasons for its inclusion here. First, it has been demonstrated that there is an interplay between the gut microbiome and epigenetics. Research suggests that gut microbiome metabolites are crucial epigenetic regulators of the host genome. They can induce epigenetic changes in key human genes and ultimately lead to the development of disease (Yuille et al, 2018). For example, changes in the diet seem to influence microbiome composition and affect the regulation of histone methylation and demethylation in the host genome (Krautkramer et al, 2016). A more specific but relatively recent area of research explores the ways in which epigenetic alterations brought about by the microbiome influence the development of cognitive function and behaviour and the development of neuropsychiatric disorders. Even though the specific underlying mechanisms are not yet fully understood, promising strands of research suggest that changes in the microbiome alter neuroactive signals via the vagus nerve, thus bringing about epigenetic changes in the brain (Kaur et al, 2021). It has also been suggested that neuroepigenetic modifications can occur due to production of short-chain fatty acids by the microbiome. These modifications underlie the pathogenesis of many neuropsychiatric conditions via inhibition of histone deacetylases (Peedicayil and Santhosh, 2021). Hence, if we are asking questions about responsibility in epigenetics, it may be helpful to consider these together with questions related to responsibility in the context of the microbiome.

While epigenetics has challenged the mechanistic view of organisms as primarily built up from genetic blueprints, microbiome studies take this knowledge one step further. Understanding epigenetics shows that we are intertwined with the environment (inside the body and outside) down to the molecular level. Gene expression is as relevant for health and disease as the information in the genes themselves. Furthermore, it has now become clear that our health is closely linked with the microbiome that is found in our gut and elsewhere. Just as with epigenetics, the composition of the gut microbiome is influenced by specific features and circumstances of the host. It has been argued that each individual's microbiome acts as a unique fingerprint (Franzosa et al, 2015). This claim is based on a growing body of literature that explores the influence of several host factors on microbiome composition, such as early-life stressors, mode of birth, diet, lifestyle, the surrounding environment and previous diseases and medications (Dong and Gupta, 2019). The link between lifestyle and the gut microbiome raises complex questions regarding responsibility for one's health.

The gut microbiome modulates the central nervous system via multiple signalling pathways that involve immune, endocrine and neural communications (Fülling et al, 2019). In addition, the microbiome has been recognized as a key regulator of the gut–brain axis. Even though the specific mechanisms of this regulation are not yet known, preclinical and clinical research supports evidence of bidirectional communication between the gut microbiome and the brain. Such communication connects the cognitive and emotional brain centres with gut functions. Hence, the term 'gut–brain axis' has been expanded to 'microbiota–gut–brain (MGB) axis' (Mayer et al, 2014).

Moreover, as with epigenetics, research into the microbiome seems to confirm the plasticity, historicity and environmentality of humans and other organisms (Meloni, 2018; Formosinho et al, 2022). At this point in this volume, we may even wonder whether maintaining sharp boundaries between areas such as genetics, epigenetics, proteomics and microbiome studies makes sense. It may be more appropriate to study the interactions and even entanglements between these domains rather than seeing them as separate. This may hold true for work in both science and ethics.

In what follows, we first discuss the implications of research into the microbiome for what it means to be human. Starting from a reflection on scientific findings on the relationship between the microbiome and mental states, we discuss what the fact that the human 'self' seems to coexist with trillions of microbes means for an understanding of responsibility. In the second section, we turn the question around, and discuss the implications for responsibility of considering humans as holobionts. We also discuss the implications of what it means to share our microbiome with non-human beings who are close to us. Finally, we reflect on what these findings imply for questions about ethics and responsibility.

Impact of knowledge about the microbiome on what it means to be human

A healthy gut and a healthy mind?

The fact that new information regarding the microbiome may challenge an atomistic view of human beings, just like epigenetics does, seems straightforward. But what does it mean that we are entangled not just with the external environment but also inhabited by and maybe even governed by other creatures? Let us look at the relationship between the microbiome and the human mind. The gut–brain axis is a focus of much present-day research. Indeed, microbiome studies are often undertaken as part of biomedical research to investigate the relationship between microbes and health status or mental wellbeing. We first discuss some of these findings.

Use of probiotics, antibiotics or faecal transplants to manipulate the commensal gut microbiota has been found to influence the behaviour of rats. These findings support the evidence that gut bacteria influence brain processes (Cryan and O'Mahony, 2011). Use of germ-free rats and mice has also enabled researchers to investigate how the gut microbiome influences the animals' behaviour (Cryan and O'Mahony, 2011). These advances in microbiomics have led to a better understanding of microbiome–host interactions. Moreover, many studies over the last decade have led to an increasing recognition of the role of the microbiome in the development of neuropsychiatric conditions in humans, including anxiety, depression and schizophrenia (Grochowska et al, 2018). These conditions are often associated with a 'dysbiotic' microbiome. Dysbiosis is an imbalance in the gut microbiome composition that favours the abundance of proinflammatory and pathogenic species and decreases microbiome diversity (Floch et al, 2017). Some studies have suggested using microbiome compositions as biomarkers for neuropsychiatric conditions (Zhu et al, 2020). However, many philosophical and scientific questions remain. Autistic people, for example, often experience difficulties in eating, which in itself will influence the microbiome's composition. Hence, the idea that there is a one-way causal pathway by which microbes determine our mental states seems naive. After all, how we feel or how we experience the world is highly likely to influence what we prefer to eat and hence the composition of the microbiome as well.

Perhaps research on the microbiome challenges simple causal explanations in psychiatric and developmental conditions even more than findings in epigenetics do. Those responsible for science communication should not shy away from conveying these complexities to the general public. It also means that responsibility for one's own or other people's mental health is more complex than what certain lifestyle coaches may suggest. Indeed, research in epigenetics and in microbiomics gives us powerful reasons to adopt a complex view of mental states, with feedback loops and sensitivity to external influences, genes, and what goes on in our gut. Moreover, we may question the concept of a 'normal' versus a dysbiotic microbiome. Just as others have argued in the case of epigenetics, the dynamic and reactive nature of the microbiome may challenge the notion of normality here (Dupras and Ravitsky, 2016).

The French philosopher and medical doctor Georges Canguilhem questioned the idea that pathology can be measured by looking at the body's internal states alone (Canguilhem, 1989). Pathology, according to Canguilhem, arises if there is a mismatch between organism and milieu to such an extent that the organism cannot proactively adapt its environment and itself to suit its needs. In this respect, we may challenge the idea of an 'abnormal' microbiome. This challenge becomes still more pressing if we look at the relationship between the microbiome and certain psychiatric conditions. It has been argued by philosophers and psychiatrists alike that to rely on concepts of normality to compare neurodivergent with neurotypical people is problematic. There may not be such a thing as an unhealthy microbiome, only one that is not adapted to its current environment. This is similar to the mismatch hypothesis in epigenetics, which suggests that there is no absolute way of defining a good or bad epigenome, just one that is adaptive or mal-adaptive in a specific environment. At the same time, the propitious findings of microbiome research, such as the correlation of certain microbiomic states with psychiatric diagnoses and the importance of the microbiome in shaping the immune system (Postler and Ghosh, 2017), show how our minds are closely interlaced with the micro-organisms inhabiting our guts.

The microbiome and the human self

In Chapter 1, Kristien Hens argues that epigenetics throws the enlightenment idea of the atomistic and autonomous individual down the drain, as it demonstrates, at a molecular level, that organisms are deeply entangled with their environment. Indeed, (primarily Western) conceptual certainties about the biological basis of our identity and sense of self are being challenged in light of post-genomic research findings. Microbiome research, as a post-genomic science, may even have more profound ontological implications (Suárez and Triviño, 2019). After all, what is a human being if it not only carries trillions of other organisms but is primarily composed of these organisms and even influenced by them when it comes to personality, responsibility and wellbeing? The gut microbiome is an indispensable component of the physiological functioning of the host. In Chapter 4, Anna Smajdor shows how epigenetics challenges the idea that what primarily matters for identity is an individual's unique combination of genes. Given that an organism's microbiome is unique and influences phenomena that we usually associate with identity, such as personality, it adds a further element to the question of identity and uniqueness.

For example, there are approximately as many microbial cells as human cells in the body (Sender et al, 2016). These findings have implications for our understanding of human identity. Natural sciences have traditionally relied on a biological view in which the human genome, brain and adaptive immune system constitute an individual self (Rees et al, 2018). However, what makes human beings human is something that philosophers have tried to answer for a long time. We suspect that 'half microbial' is more than they bargained for. We can circumvent the implications of the above findings by no longer seeing these microbes as separate, external organisms that are part of the environment that just happens to be in our gut. Instead, we could consider them as part of us. After all, there are many examples in the history of life itself of unicellular organisms merging with other organisms, mitochondria being a famous example.

We may wonder, however, why we as philosophers are so hung up on personal identity and the human as a discrete and well-circumscribed entity. Personal identity has been linked to numerical identity: numerically, human beings remain one and the same over time. However, research into the microbiome may support the view that human beings are more 'ship of Theseus-like' than we usually think. Like the ship of Theseus, all of a human being's material components, such as cells and also microbes, are gradually replaced throughout one's lifetime. At the same time, personal identity has been firmly linked to the unique set of genes that we acquire at conception, with the exception of the case of monozygotic twins. This has influenced how we think about responsibility towards future people, Derek Parfit's non-identity problem being a case in point (see Chapter 4). Indeed, giving up on a fixed sense of identity seems dangerous, and risks undermining the fundamental prerequisites of morality. Questions of moral responsibility seem to imply a specific answer to the question who is responsible for whom. At the same time, we may also ask ourselves whether we are missing opportunities to think about responsibility differently if it is so firmly linked to a numerical or genetic interpretation. After all, social sciences and humanities have investigated personal identity by looking at the various ways people have understood themselves. Individuality may be understood as inherent in the continuity of a person's past, present and future (Rex and Mason, 1986). The fact that our 'self' is partly microbial may not conflict as much with the narrative self as with a personal identity based on old-fashioned ideas about biology and genes.

Microbes and environmental entanglements

The examples given above suggest that the relationship humans have with their microbiome is one of symbiosis. Not only do gut bacteria help us digest food, but they are also tightly intertwined with our personality, to the point that the gut has been called 'the second brain'. This collective interaction between the host and microbiome has been called the 'holobiont'. According to Bosch and Miller (2016, p 1), a holobiont is 'an association comprised of the macroscopic host and synergistic interdependence with bacteria, archaea, fungi, and numerous other microbial and eukaryotic species'. We may conceive of the microbiome as an environmental factor that influences us, while, at the same time, we are our microbiome's environment. However, even when thinking about epigenetics, the concept of environment is problematic. Epigenetics is often conceptualized as the molecular 'link' between genes and the environment. Such conceptualization assumes that, as causal agents, genes and environment operate at the same level. For example, a health problem may be due to the environment, genes or a combination thereof. This distinction does normative work: if one's health problem is caused by an environmental factor, it is often assumed that one's responsibility to do something about it is greater than in the case of a genetic cause. Nevertheless, the environment is many things. It is the intracellular environment, the environment within the body, the local environment, or even something beyond. We may see the workings of organisms not as 'genes versus the rest' but as a dynamic and interacting system in which genes are but one aspect. The microbiome challenges the dichotomy of genes versus environment even more: now, a factor *within* the body is added. At the same time, the concept of the environment may in itself have the connotation of being external and fixed, as Formosinho and colleagues have argued in their excellent paper (Formosinho et al, 2022). They state that 'microbiome research introduces challenges regarding usage of the term environment: what constitutes an environment, for whom, and with which consequences for health?' (Formosinho et al, 2022, p 148) They situate microbiome research in 'a history of reaching for a more scientific medicine; a more controlled, precise and generalizable knowledge that would separate the body from the environment and locate it instead in the aperspectival "view from nowhere" of the clinic' (Formosinho et al, 2022, p 152). The microbiome further complicates this aperspectival view. Formosinho and colleagues propose 'environmentality' as 'the state or quality of being a causal context for something else', a 'firmly perspectival concept aware of its own situatedness and the situatedness of its object of study' (Formosinho et al, 2022, p 152). For them, environmentality is an epistemic tool that has 'helped us trace lines of relationality across scales, back in time, through flesh and across organismic boundaries' (Formosinho et al, 2022, p 155). As such, the concept also seems suitable as a way of looking at epigenetic research.

At the same time, microbiome findings raise questions about the status of human beings as holobionts, the status of human beings in general, and even what it means to belong to a particular species. It has become apparent that, through epigenetics and the microbiome, human beings are deeply entangled with the environment inside and outside their skin. Given the link between the gut microbiome and our brain, it is tempting to assume that, although we are partly microbes, at least the specific ecosystem of microbes with which we are populated must surely be distinctly human. However, studies have shown that owners of companion animals such as dogs, share gut microbiomes with their pets (Coelho et al, 2018). Moreover, a recent study has suggested that urban populations of coyotes, crested anole lizards and white-crowned sparrows share more similar gut microbiota with humans than with non-urban members of their own species (Dillard et al, 2022). We discuss what this might mean later on.

Microbes, ethics and responsibility

Contributors to this volume have discussed a number of important questions that epigenetics may raise concerning individual, collective or parental responsibility. As shown above, findings generated by research into the microbiome may raise similar questions. For example, we could ask what the implications of these findings are for parental responsibility. Should we conclude that parents have an even bigger responsibility to feed their children healthy food if this gives them a 'healthy microbiome'? Should Caesarean sections only be seen as a last resort because of the importance of vaginal microbiome transmission to babies (Hoang et al, 2021)? As many have argued with regard to epigenetics, such conclusions tend to neglect the contexts that influence people's decisions or limit their opportunities to make decisions at all.

Moreover, just as with epigenetics, the link between the microbiome and health and disease is complex. A healthy microbiome may be regarded as such more because it adapts effectively to particular environments than by virtue of its intrinsic properties. This insight may suggest that the balance tips to more collective and forward-looking responsibility (see also Chapter 2 for a discussion of those concepts in an epigenetics context). Such collective forward-looking responsibility then implies the need to ensure that the environment we have is one in which organisms can flourish. In what follows, we do not provide an overview of all responsibility issues that may arise with the increasing knowledge about the microbiome. Instead, we focus on two domains. First, we hint at some points to consider for medical ethics. Second, we argue that microbiome research is yet another proof that medical ethics, or bioethics, and environmental ethics, should not be seen as separate endeavours, but should always be undertaken in synergy.

The microbiome and medical ethics

Medical ethicists who have considered ethical issues related to the genome have often covered issues related to privacy, confidentiality, consent and the return of test results. Similar issues may arise when considering epigenetic and microbiome research. For example, epigenetic and microbiome data may contain sensitive information that could identify the donor. Moreover, they may contain not only genetic, but also phenotypic, information, which may be even more interesting for insurance companies. They may contain more relevant information about the health of the subject in question than mere genetic data. Genetic data are currently governed by strict regulations. At the same time, stool samples are considered waste and are not subject to the same scrutiny. Given the increasing knowledge in the field of the human microbiome, it may be at least as necessary to reflect on responsible management of stool samples as on the management of genetic biobanks.

In addition to research on stool samples, treatments such as faecal transplants to treat metabolic diseases or even psychiatric conditions should be approached with due care. The impact of mental health issues on people's wellbeing is the focus of significant attention. Being able to manipulate the gut microbiome and hence influence one's mental health seems promising in that light. However, there are several reasons why one should tread carefully. Microbiomegutbrain findings are often interpreted as simple biological explanations for psychiatric conditions but the truth may well be more complex.

Microbiome explanations for psychiatric conditions may add to the idea that 'it is all in your biology'. Biological explanations of mental health issues can be liberating as they sometimes relieve sufferers from feeling guilty. Indeed, meta-analytical evidence has shown that, when the public accepts biological explanations of mental disorders, this may reduce the moral responsibility attributed to sufferers by revealing that mental health problems are not the result of bad character or weakness (Schomerus et al, 2012). A more recent meta-analysis has shown that neurobiological explanations that conceptualize a psychiatric disorder as a brain disease tend to have stigmatizing consequences (Loughman and Haslam, 2018), provoking fear and a desire to maintain social distance. For example, people assume that neurobiological explanations imply that the affected person cannot control their actions. As the brain is perceived as the source of free will and actions, any explanation that casts doubt on its integrity risks being understood as an indication that the affected person is unpredictable and potentially dangerous (Loughman and Haslam, 2018).

Having your mental health issues straightforwardly explained through your gut microbiome may have similar effects. At the same time, the nature of the microbiome itself challenges such simplistic causal attributions to 'biology'. Indeed, viewing mental states through a microbial lens diminishes the lines between body and mind and physical and mental health. It undermines a reductionist conception of mental health and disease and introduces a new conception that combines biological, genetic, social and environmental factors. Moreover, the microbiome's dynamic nature may emerge as a more fitting approach to studying mental conditions as it leaves more room for understanding personal and environmental circumstances synchronously with biological factors (Ahmed and Hens, 2021). In this sense, a responsible application of microbiome findings in mental health practice necessarily implies a holistic approach to mental health. Science communicators and the media have work to do to ensure that the complexity of the links between mental health, the brain and the gut is clear, rather than presenting the microbiome as a direct cause for certain issues. Given the still high prevalence of 'gene for psychiatric condition X' language in the popular media, decades after the idea of a gene for a condition has been debunked, there is still a long way to go.

In addition to the danger of continuing a reductionist vision of the causes of mental health issues, there may be issues related to specific treatments. In the first part of this chapter, we challenged the idea of a 'numerical' or 'biological' identity in favour of a narrative one. That does not mean that the former is necessarily less 'real' or more fluid, or that interruptions in biology are not disruptive to that narrative. For example, suppose specific microbiome treatments such as a faecal transplant from a donor without a mental illness could treat mental illness in the recipient. Just like technical approaches such as deep brain intervention, these treatments may profoundly affect how one sees oneself and one's life story. Therefore, applying such interventions responsibly requires paying attention to the stories people tell about themselves and their afflictions. Such treatments should hence also be part of a holistic approach.

Microbes and bioethics as global bioethics

We ended the last paragraph with a plea for a holistic approach to microbiome science and treatment: microbiomes should not be seen as simple causal agents of disease, nor should they solely be targeted as simple biological solutions. The nature of our relationship with the microbiome suggests the need for an appreciation of the complexity of an organism's functioning, the entanglement of human beings and microbiome – if it even makes sense to make that distinction – and of holobiont and the outer environment. Just as epigenetics may be characterized as the molecular link between genes and environment, a link that may even render the distinction between the two obsolete, the microbiome proves that humans and other organisms are not 'standalone' beings. They are composed of various types of cooperating life, which, in itself, is influenced by the immediate environment of the human gut and the human diet, and also by the external environment. Just as with epigenetics, this gives us reason to reassess the scope of the discipline of bioethics.

Nowadays, bioethics is often seen as distinct from environmental ethics. Given the recent findings regarding health and the environment, we may wonder whether this distinction is sustainable or helpful or even a responsible way of practicing ethics. Epigenetics and studies into the microbiome may be a wake-up call regarding what bioethics should be about. We may need global bioethics, to use the words of Van Rensselaer Potter, who mourned that, in the 1970s, bioethics had been reduced to medical ethics (Potter 1988). If we assume that the task of bioethics is to ensure that 'good' is done in medical practice, it may be nonsensical to neglect the broader context in which the good is to be done. This broader context necessarily involves thinking about environmental issues. Moreover, environmental ethics itself has much to gain from looking at microbiome findings.

Discussions about anthropocentric approaches versus ecocentric approaches may become meaningless in the light of the knowledge that we are the environment, and the environment is us. As seen before, an example of this is the finding that the microbiomes of all urban dwellers, be they humans, coyotes or lizards, share more features in common than they do with their non-urban cousins (Dillard et al, 2022). We could take this as a call for a stricter separation between urban humans and wild animals. An all-too-easy interpretation would be that we have colonized these animals with our microbiome. However, these findings could also inspire us to think differently about species boundaries, locality and kinship. In this respect, rather than trying to lay down strict boundaries between ourselves and non-humans, between domestic and wild animals, between society and nature, we could acknowledge that we, humans, coyotes and microbes, are all in this together. Human and non-human health are not necessarily different spheres. We are responsible for ensuring a liveable future for all.

In her brilliant book *Philosophy of Microbiology*, Maureen O'Malley argues that, rather than focusing on plants and (specifically human) animals, philosophers of biology should instead think from the starting point of microbial life (O'Malley, 2014). She points out how cooperation, symbiosis and entanglement have been built into life for billions of years. We conclude with the suggestion that it is not only philosophers of biology who should start from microbial life, but bioethicists too, who have much to gain from 'thinking with' the human as holobiont.

Contributor statement

E.A. and K.H. both contributed to the structure and content of this chapter. E.A. and K.H. both gave feedback on and edited the whole manuscript and agree with this final version.

References

Ahmed, E. and Hens, K. (2021) 'Microbiome in precision psychiatry: an overview of the ethical challenges regarding microbiome big data and microbiome-based interventions, *AJOB Neuroscience*, 13(4): 270–86. https://doi.org/10.1080/21507740.2021.1958096.

Bosch, T.C.G. and Miller, D.J. (2016) *The Holobiont Imperative: Perspectives from Early Emerging Animals*, Berlin: Springer.

Canguilhem, G. (1989) *The Normal and the Pathological*, New York: Zone Books.

Coelho, L.P., Kultima, J.R., Costea, P.I., Fournier, C., Pan, Y., Czarnecki-Maulden, G., et al (2018) 'Similarity of the dog and human gut microbiomes in gene content and response to diet', *Microbiome*, 6(1): 72. https://doi.org/10.1186/s40168-018-0450-3.

Cryan, J.F. and Dinan, T.G. (2012) 'Mind-altering microorganisms: the impact of the gut microbiota on brain and behaviour', *Nature Reviews Neuroscience,* 13(10): 701–12. https://doi.org/10.1038/nrn3346.

Cryan, J.F. and O'Mahony, S.M. (2011) 'The microbiome–gut–brain axis: from bowel to behavior', *Neurogastroenterology & Motility*, 23(3): 187–92. https://doi.org/10.1111/j.1365-2982.2010.01664.x.

Dillard, B.A., Chung, A.K., Gunderson, A.R., Campbell-Staton, S.C. and Moeller, A.H. (2022) 'Humanization of wildlife gut microbiota in urban environments', *eLife*, 11: e76381. https://doi.org/10.7554/eLife.76381.

Dong, T.S. and Gupta, A. (2019) 'Influence of early life, diet, and the environment on the microbiome', *Clinical Gastroenterology and Hepatology*, 17(2): 231–42. https://doi.org/10.1016/j.cgh.2018.08.067.

Dupras, C. and Ravitsky, V. (2016) 'The ambiguous nature of epigenetic responsibility', *Journal of Medical Ethics*, 42(8): 534–41.

Floch, M.H. (2018) 'The role of prebiotics and probiotics in gastrointestinal disease', *Gastroenterology Clinics of North America*, 47(1): 179–91.

Formosinho, J., Bencard, A. and Whiteley, L. (2022) 'Environmentality in biomedicine: microbiome research and the perspectival body', *Studies in History and Philosophy of Science*, 91: 148–58. https://doi.org/10.1016/j.shpsa.2021.11.005.

Franzosa, E.A., Huang, K., Meadow, J.F., Gevers, D., Lemon, K.P., Bohannan, B.J.M., et al (2015) 'Identifying personal microbiomes using metagenomic codes', *Proceedings of the National Academy of Sciences USA*, 112(22): E2930–8. https://doi.org/10.1073/pnas.1423854112.

Fülling, C., Dinan, T.G. and Cryan, J.F. (2019) 'Gut microbe to brain signaling: what happens in vagus …', *Neuron*, 101(6): 998–1002. https://doi.org/10.1016/j.neuron.2019.02.008.

Grochowska, M., Wojnar, M. and Radkowski, M. (2018) 'The gut microbiota in neuropsychiatric disorders', *Acta Neurobiologiae Experimentalis*, 78(2): 69–81.

Hoang, D.M., Levy, E.I. and Vandenplas, Y. (2021) 'The impact of Caesarean section on the infant gut microbiome', *Acta Paediatrica*, 110(1): 60–7. https://doi.org/10.1111/apa.15501.

Kaur, H., Singh, Y., Singh, S. and Singh, R.B. (2021) 'Gut microbiome-mediated epigenetic regulation of brain disorder and application of machine learning for multi-omics data analysis', *Genome*, 64(4): 355–71. https://doi.org/10.1139/gen-2020-0136.

Krautkramer, K.A., Kreznar, J.H., Romano, K.A., Vivas, E.I., Barrett-Wilt, G.A., Rabaglia, M.E., et al (2016) 'Diet–microbiota interactions mediate global epigenetic programming in multiple host tissues', *Molecular Cell*. 64(5): 982–92. https://doi.org/10.1016/j.molcel.2016.10.025.

Loughman, A. and Haslam, N. (2018) 'Neuroscientific explanations and the stigma of mental disorder: a meta-analytic study', *Cognitive Research: Principles and Implications*, 3(1): 43. https://doi.org/10.1186/s41235-018-0136-1.

Mayer, E.A., Knight, R., Mazmanian, S.K., Cryan, J.F. and Tillisch, K. (2014) 'Gut microbes and the brain: paradigm shift in neuroscience', *Journal of Neuroscience*, 34(46): 15490–6.

Meisel, J.S. and Grice, E.A. (2017) 'The human microbiome', in G.S. Ginsburg and H.F. Willard (eds), *Genomic and Precision Medicine. Foundations, Translation and Implementation* (3rd edn), Cambridge, MA: Academic Press, pp 3–77. https://doi.org/10.1016/b978-0-12-800681-8.00004-9.

Meloni, M. (2018) 'A postgenomic body: histories, genealogy, politics', *Body & Society*, 24(3): 3–38. https://doi.org/10.1177/1357034X18785445.

Ochoa-Repáraz, J. and Kasper L.H. (2016) 'The second brain: is the gut microbiota a link between obesity and central nervous system disorders?', *Current Obesity Reports*, 5(1): 51–64. Doi: 10.1007/s13679-016-0191-1.

O'Malley, M. (2014) *Philosophy of Microbiology*, Cambridge: Cambridge University Press.

Peedicayil, J. and Santhosh, S. (2021) 'Epigenetic aspects of the microbiota and psychiatric disorders', in J. Peedicayil, D.R. Grayson and D. Avramopoulos (eds), *Epigenetics in Psychiatry* (2nd edn), Amsterdam: Elsevier, pp 783–791.

Postler, T.S. and Ghosh, S. (2017) 'Understanding the holobiont: how microbial metabolites affect human health and shape the immune system', *Cell Metabolism*, 26(1): 110–30. https://doi.org/10.1016/j.cmet.2017.05.008.

Potter, V.R. (1988) *Global Bioethics: Building on the Leopold Legacy*, East Lansing: Michigan State University Press.

Rees, T., Bosch, T. and Douglas, A.E. (2018) 'How the microbiome challenges our concept of self', *PLoS Biology*, 16(2): e2005358. https://doi.org/10.1371/journal.pbio.2005358.

Rex, J. and Mason, D. (eds) (1986) *Theories of Race and Ethnic Relations* (1st edn), Cambridge: Cambridge University Press.

Schomerus, G., Schwahn, C., Holzinger, A., Corrigan, P.W., Grabe, H.J., Carta, M.G., et al (2012) 'Evolution of public attitudes about mental illness: a systematic review and meta-analysis', *Acta Psychiatrica Scandinavica*, 125(6): 440–52. https://doi.org/10.1111/j.1600-0447.2012.01826.x.

Sender, R., Fuchs, S. and Milo, R. (2016) 'Revised estimates for the number of human and bacteria cells in the body', *PLoS Biology*, 14(8): e1002533.

Suárez, J. and Triviño, V. (2019) 'A metaphysical approach to holobiont individuality: holobionts as emergent individuals', *Quaderns de Filosofia*, 6(1): 59–76. https://doi.org/10.7203/qfia.6.1.14825.

Wu, Z.A. and Wang, H.X. (2019) 'A systematic review of the interaction between gut microbiota and host health from a symbiotic perspective', *SN Comprehensive Clinical Medicine*, 1(3): 224–35. https://doi.org/10.1007/s42399-018-0033-4.

Yuille, S., Reichardt, N., Panda, S., Dunbar, H. and Mulder, I.E. (2018) 'Human gut bacteria as potent class I histone deacetylase inhibitors *in vitro* through production of butyric acid and valeric acid', *PLoS One*, 13(7): e0201073. https://doi.org/10.1371/journal.pone.0201073.

Zhu, S., Jiang, Y., Xu, K., Cui, M., Ye, W., Zhao, G., et al (2020) 'The progress of gut microbiome research related to brain disorders', *Journal of Neuroinflammation*, 17(1): 25. https://doi.org/10.1186/s12974-020-1705-z.

Index

References to endnotes show both the page number and the note number (15n1).